Differential Calculus

A. Avez

Translated by
D. Edmunds

DOVER PUBLICATIONS, INC.
Mineola, New York

Bibliographical Note

This Dover edition, first published in 2020, is an unabridged republication of the work as published in 1986 by John Wiley & Sons, Inc., New York, as a Wiley-Interscience Publication. It was originally published in French under the title *Calcul Differentiel* by Masson, Paris, in 1983.

Library of Congress Cataloging-in-Publication Data

Names: Avez, A. (André), author.
Title: Differential calculus / A. Avez.
Description: Mineola, New York : Dover Publications, Inc., 2020. | Includes bibliographical references and index. | Summary: "Original, rigorous, and lively; this text offers a concise approach to classical and contemporary topics in differential calculus. The subject is presented in a Banach space setting, covering vector fields, one-parameter groups of diffeomorphisms, the Morse-Palais lemma, and differentiable submanifolds. Suitable for upper-level baccalaureate and graduate students of analysis, the treatment also examines applications to differential equations and the calculus of variables"—Provided by publisher.
Identifiers: LCCN 2020002174 | ISBN 9780486845647 (trade paperback) | ISBN 0486845648 (trade paperback)
Subjects: LCSH: Differential calculus.
Classification: LCC QA304 .A85 2020 | DDC 515/.33—dc23
LC record available at https://lccn.loc.gov/2020002174

Manufactured in the United States by LSC Communications
84564801
www.doverpublications.com

2 4 6 8 10 9 7 5 3 1

2020

To the memory of Renée Gallai and Gaston Roux, to Jean Houlle. Three people who taught me most of the useful things that I know. Two scholars the like of whom we shall not see again.

Contents

Foreword

Several years ago the course committee of the Université Pierre et Marie Curie laid down a minimum programme for the unit of instruction of differential calculus:

The notion of derivative; C^r and C^∞ classes
Symmetry of the second derivative
The Taylor formula in several variables
Implicit functions; the case of constant rank
Differential equations: the theorem of existence and dependence on initial conditions and upon parameters, in the lipschitzian case.

This book covers the above programme, and a little more.

The appendices are of two kinds. Some recall concepts which are, strictly speaking, outside differential calculus and which are included to make the text self-contained, while the others deal with topics which may be omitted at a first reading.

Bibliographical references are given in full detail in the body of the text if they are used only once. If not, the author's name is given (rather than a number), and the alphabetical list at the end of the book gives the corresponding work.

While this book can be used in connection with study for the Differential Calculus certificate, it is intended to be an introduction to more advanced works, such as that by R. Abraham and J. Marsden, to which it owes a great deal.

Finally, the author wishes to thank Professor Paul Malliavin, who knew how to overcome the author's laziness, and without whom this book would never have seen the light of day. He also thanks Professor B. El Mabsout, who read the manuscript and prevented certain errors.

The author also apologizes to his wife and sons for having deprived them of time which normally would have been devoted to family life.

Notation

$\mathbf{R}^+$:	set of positive real numbers
$[a, b]$:	set of real numbers t such that $a \leqslant t \leqslant b$
$]a, b[$:	set of real numbers t such that $a < t < b$
sup:	least upper bound
$\max\{a, \ldots\}$:	the greatest of the numbers $a, \ldots$
f^{-1}:	inverse (or reciprocal) of the bijection f
id_E:	identity map of E to E
$\bar{A}$:	closure of the set A
$\exp(t)$:	exponential of t
ch:	hyperbolic cosine
sh:	hyperbolic sine
v.s.	vector space
E^*:	dual of the vector space E
E^c:	complexification of the real vector space E
$\mathscr{L}(E, F)$:	set of all continuous linear maps of a vector space E to a vector space F
$\mathrm{End}(E)$:	(E, E)
$\mathrm{GL}(E)$:	group of continuous linear bijections of a vector space E onto itself
$\mathrm{SL}(n)$:	unimodular group of $\mathbf{R}^n$
$E \oplus F$:	direct sum of vector spaces E and F
$\mathrm{Ker}(f)$:	kernel of a linear map f
$\det(f)$:	determinant of a linear endomorphism f
$\mathrm{tr}(f)$:	trace of a linear endomorphism f
^tf:	transpose of a linear map f
$\lVert x \rVert_V$:	norm in a normed linear space V
$B(a, r)$ or $B_r(a)$:	open ball with centre a and radius r
$\langle\, ,\, \rangle$:	scalar or hermitian product
f^*:	adjoint of a linear map f
S^n:	n-dimensional unit sphere
$\mathrm{O}(n)$:	orthogonal group of $\mathbf{R}^n$
$Df(a)$:	derivative of f at a
Df:	derivative map of f

f':	derivative of the function f
$T_a f$:	tangent map of f at a
grad f:	gradient of the function f
$D_r f$:	partial derivative with respect to the rth variable
C^k:	class C^k
$C^k(U; F)$:	set of all maps of class C^k from U to F
$\lVert f \rVert_{C^r}$:	C^r norm of f
$\mathrm{Diff}^r(E)$:	group of all C^r diffeomorphisms from E to E
$\phi_* X$:	image of the vector field X under the derivative map ϕ
$T_m M$:	tangent space at m to the sub-manifold M

1

The concept of a derivative

This chapter is devoted to the definition of a derivative, and to those of its elementary properties which make *no* use of the ideas of a complete space or of integration.

PRELIMINARIES

In this chapter the vector spaces are normed spaces E over the field $K = \mathbf{R}$ or $\mathbf{C}$. Appendix A recalls some of the properties of these spaces which are used here. In particular the norm of E will be denoted by $\|\cdot\|_E$, or by $\|\cdot\|$ if there is no ambiguity.

When several spaces occur in the same statement, it is understood that they are all over the same field.

1.1 DEFINITIONS AND EXAMPLES

A function f: $\mathbf{R} \to \mathbf{R}$ is differentiable at $a \in \mathbf{R}$ if there is a real number $f'(a)$ such that $\lim_{h \to 0} [f(a+h) - f(a)]/h = f'(a)$. This definition makes no sense for a map f: $\mathbf{R}^n \to \mathbf{R}^m$, $n > 1$ (how can we divide by a vector h of $\mathbf{R}^n$?); but it may be reformulated thus: $\lim_{h \to 0} |f(a+h) - f(a) - f'(a)h| / |h| = 0$, and this gives rise to the following generalization.

Definition 1.1

Let U be a non-empty open subset of a normed v.s. E and let f be a map of U to a normed v.s. F. We say that f is differentiable at $a \in U$ if there is a *continuous* linear map $L \in \mathscr{L}(E;F)$ such that $\lim_{h \to 0} \|f(a+h) - f(a) - Lh\|_F / \|h\|_E = 0$. This assumes that $\|h\|$ is so small that $a + h \in U$.

Let $k \in E$ be arbitrarily given, and let t be real. Since U is *open*, $a + tk \in U$ if $|t|$ is small enough. Replacing h by tk in the definition, we see that if L exists it is unique and is given by $L(k) = \lim_{t \to 0} [f(a+tk) - f(a)]/t$. It is simply to ensure the uniqueness of L that U is assumed to be open.

We call L the derivative of f at a and denote it by $Df(a)$; it is an element of $\mathscr{L}(E; F)$.

$Df(a)k$ is called the derivative of f in the direction of the vector k. We have

$$Df(a)k = \frac{\mathrm{d}}{\mathrm{d}t} f(a+tk)\,|_{t=0} \tag{1.1}$$

where the right-hand side is the usual derivative of $t \in \mathbf{R} \mapsto f(a+tk) \in F$ at $t=0$.

Recall that a real-valued function g of a real variable t, defined on a neighbourhood of zero, is said to be $\mathrm{o}(t)$ if $\lim_{t\to 0} g(t)/t = 0$. Thus $f(a+h) = f(a) + Df(a)h + \mathrm{o}(\| h \|)$, and $f(a) + Df(a)h$ is seen to be an approximation to f of the first order in h in the neighbourhood of a.

Remarks

(a) The definition of the derivative depends on the norms of E and F. Let us replace them by equivalent norms (see §A.2.4). U remains open since the topologies of E and F are unchanged. It is easy to see that f remains differentiable and that its derivative at a is still $Df(a)$. If, in particular, E and F are finite-dimensional, differentiability does not depend on the norm, for they are all equivalent (§A.2.5), and the hypothesis of continuity of $Df(a)$ is redundant, for every linear map is continuous (§A.2.5).

(b) We known that if $E = K$, $\mathscr{L}(E; F)$ may be identified with F by the canonical isomorphism $L \in \mathscr{L}(E; F) \mapsto L.1 \in F$ (Appendix, Theorem A.8). Thus $Df(a)$ is identified with $Df(a).1$, which is simply the usual derivative $f'(a)$ since

$$Df(a).1 = \frac{\mathrm{d}}{\mathrm{d}t} f(a+t)\,|_{t=0} = f'(a)$$

(c) If $F = K$, $Df(a)$ is a continuous linear functional. If, in addition, the norm of E arises from an inner product $\langle , \rangle$, then $E^* = \mathscr{L}(E; K)$ may be identified with E by the canonical isomorphism $x \in E \mapsto \langle x, \cdot \rangle \in E^*$, and $Df(a)$ is the image of a vector grad $f(a)$, called the gradient of f at a and characterized by $\langle \operatorname{grad} f(a), v \rangle = Df(a)v$ for all v in E. Note that grad $f(a)$ depends on the inner product chosen, while $Df(a)$ does not.

(d) *Jacobian matrix.* If $E = K^n$, $F = K^m$, the matrix of the linear map $Df(a)$ in the canonical bases is a matrix with m rows and n columns. It is called the Jacobian matrix of f at a. We shall learn later how to determine its elements.

(e) A map which is differentiable at a is continuous at a since, as $Df(a)$ is continuous, $\lim_{h\to 0} f(a+h) = f(a)$. (The converse is false: $t \in R \mapsto |t|$ for $t = 0$).

(f) The notion of derivative may be extended to affine spaces, and this possibility is of great importance in physics and mechanics.

Let U be an open subset of the affine space E with associated v.s.$\vec{E}$ which is normed, and let f be a map of U to an affine space F with normed associated v.s. $\vec{F}$. We say that f is differentiable at $a \in U$ if there exists $L \in \mathscr{L}(\vec{E}; \vec{F})$ such that

$$f(a+\vec{h}) = f(a) + L\vec{h} + o(\|\vec{h}\|)$$

By means of a choice of *fixed* origins in E and F this notion reduces to that for v.s., and we shall confine ourselves to this to simplify the notation (the arrows may now be omitted).

Definitions 1.2

If f: $U \to F$ is continuous at each point of the open subset U of E, we say that f is of class C^0, or is C^0, in U, and it will be convenient to set $D^0 f = f$.

If f is differentiable at each point of U, we say that f is differentiable on U. In this case, the mapping Df of the open set U to the normed v.s. $\mathscr{L}(E; F)$, defined by $x \rightsquigarrow Df(x)$, is called the derivative of f, and it will be convenient to set $D^1 f = Df$.

If Df is continuous for the topologies of U and $\mathscr{L}(E; F)$ defined by their norms, we say that f is continuously differentiable, or of class C^1, or simply C^1, on U.

If A is a subset of E, not necessarily open, we say that f: $A \to F$ is differentiable (resp. C^1) on A if f is the restriction to A of a differentiable (resp. C^1) map from an open set containing A to F.

Examples

(a) Let $E = F = R^2$, normed by $|(x, y)| = |x| + |y|$, and let f: $E \to F$ be defined by $f(x, y) = (x + y, xy)$. If $a = (a_1, a_2)$ and $h = (h_1, h_2)$, we have $f(a + h) = f(a) + (h_1 + h_2, h_1 a_2 + h_2 a_1) + (0, h_1 h_2)$. Since $\|(0, h_1 h_2)\| = |h_1 h_2| \leqslant (|h_1| + |h_2|)^2 = \|h\|^2$, then $(0, h_1 h_2) = o(\|h\|)$. Thus f is differentiable at a and $Df(a)h = (h_1 + h_2, h_1 a_2 + h_2 a_1)$; hence its Jacobian matrix at a is

$$\begin{pmatrix} 1 & 1 \\ a_2 & a_1 \end{pmatrix}$$

(b) By §A.1.3, the space $E = C^1([0, 1])$ of functions u: $[0, 1] \to \mathbf{R}$ which have a continuous derivative is normed by $\|u\|_{C^1} = \|u\|_{C^0} + \|u'\|_{C^0}$. The space $F = C^0([0, 1])$ of continuous functions v: $[0, 1] \to \mathbf{R}$ is normed by $\|v\|_{C^0}$. Let f: $E \to F$ be the mapping which to each $u \in E$ makes correspond the function $f(u)$: $t \in [0, 1] \rightsquigarrow u'(t) + tu^2(t)$. With an obvious abuse of notation, if $u, h \in E$, then $f(u + h) = f(u) + Lh + th^2$, where $Lh = h' + 2tuh$. Clearly L is linear, and it is continuous since

$$\|Lh\|_{C^0} \leqslant \|h'\|_{C^0} + 2\|u\|_{C^0}\|h\|_{C^0} \leqslant [1 + 2\|u\|_{C^0}]\,\|h\|_{C^1}$$

On the other hand, $\|th^2\|_{C^0} \leqslant (\|h\|_{C^0})^2 \leqslant (\|h\|_{C^1})^2$ and so $th^2 = o(\|h\|_{C^1})$. Thus f is differentiable on E and $Df(u)h = h' + 2tuh$.

1.2 RULES FOR CALCULATION

1.2.1 Derivative of a continuous linear map

A continuous linear map $f \in \mathscr{L}(E; F)$ is differentiable on E and its derivative $Df(a) = f$ for all $a \in E$.

For $f(a+h) = f(a) + fh$ [the $o(\|h\|)$ term reduces to zero].

1.2.2 Derivative of a continuous affine map

More generally, let $f\colon E \to F$ be a continuous affine map: $f(x) = Lx + b$, where $L \in \mathscr{L}(E; F)$, $b \in F$. Then f is differentiable on E and $Df(a) = L$.

If, in particular, f is constant its derivative is zero. We shall see that the converse is true if f is defined on a *connected* open set U. It is false if U is not connected (take $E = F = \mathbf{R}$, $U =$ disjoint union of non-empty intervals A and B, $f = 0$ on A and $f = 1$ on B).

1.2.3 Continuous bilinear maps

Let E_1, E_2 and F be three normed v.s., and let $b\colon E_1 \times E_2 \to F$ be a continuous bilinear map (§A.3). Then b is differentiable on $E_1 \times E_2$ and the value at $h = (h_1, h_2) \in E_1 \times E_2$ of its derivative at $a = (a_1, a_2)$ is $Db(a)h = b(h_1, a_2) + b(a_1, h_2)$.

In order to prove this, since $b(a+h) = b(a) + b(h_1, a_2) + b(a_1, h_2) + b(h_1, h_2)$, it is enough to show that $b(h_1, h_2) = o(\|h\|)$. But this follows from

$$\|b(h_1, h_2)\| \leqslant \|b\| \, \|h_1\| \, \|h_2\| \leqslant \|b\| \, [\|h_1\| + \|h_2\|]^2 \leqslant \|b\| \, \|h\|^2.$$

The above can be generalized. Let $E_1, \ldots, E_n$ and F be normed v.s., and let $f \in \mathscr{L}(E_1, \ldots, E_n; F)$. Then f is differentiable at every point $a = (a_1, \ldots, a_n)$ and the value at $h = (h_1, \ldots, h_n)$ of its derivative is

$$Df(a)h = \sum_{k=1}^{n} f(a_1, \ldots, a_{k-1}, h_k, a_{k+1}, \ldots, a_n).$$

1.2.4 Trace of a differentiable map

Let $f\colon E \to F$ be a differentiable map between normed v.s. A vector subspace E' of E is normed by $\|x\|_{E'} = \|x\|_E$ if $x \in E'$. Then the restriction $f|_{E'}$ of f to E' is differentiable, and its derivative is the restriction to E' of the derivative of f. The proof is obvious.

Consider a special case of the above. Suppose E is a Hilbert space and let its norm be induced by its inner product $\langle , \rangle$. If E' is a closed subspace of E, every x in E has an orthogonal projection px on E' (projection theorem). Suppose that f is a scaler valued function ($F = K$). Then the gradient of $f|_{E'}$ is the projection on E' of the gradient of f.

In fact, for every $h \in E'$ we have $\langle p \operatorname{grad} f(a), h\rangle = \langle \operatorname{grad} f(a), h\rangle = Df(a)h = Df|_{E'}(a)h = \langle \operatorname{grad} f|_{E'}(a), h\rangle$.

1.2.5 Linearity of the derivative

Let U be a non-empty open subset of a normed v.s. E, and let f and g be two maps of U to a normed v.s. F. If $k, k' \in K$, define $kf + k'g$: $U \to F$ by $(kf + k'g)(x) = kf(x) + k'g(x)$ for all $x \in U$. It is easy to see that if f and g are differentiable at $a \in U$, so is $kf + k'g$ and that

$$D(kf + k'g)(a) = kDf(a) + k'Dg(a)$$

The set of all maps which are differentiable at a (resp. on U) is thus a v.s. So is the set of all C^1 maps of U to F; we denote it by $C^1(U; F)$.

1.2.6 Derivative of a composite map (the chain rule)

Let E, F and G be three normed v.s. Let f be a map of an open subset U of E to F, and suppose that it is differentiable at $a \in U$. Let g be a map of an open subset of F, which contains $f(U)$, to G. Assume that g is differentiable at $b = f(a)$. Then $g \circ f$ is differentiable at a, and

$$D(g \circ f)(a) = Dg(f(a)) \circ Df(a)$$

where the right-hand side is the composition of the continuous linear maps $Dg(b)$ and $Df(a)$.

Proof. Set $y = f(a + h)$ and $b = f(a)$. By hypothesis we have

$$g(y) = g(b) + Dg(b)(y - b) + r(y - b)$$

where $\lim_{y \to b} \dot{r}(y - b) \| / \| y - b \| = 0$. Taking account of the linearity of $Dg(b)$ and the hypothesis $y - b = Df(a)h + s(h)$, where $\lim_{h \to 0} \| s(h) \| / \| h \| = 0$, we have

$$(g \circ f)(a + h) - (g \circ f)(a) - Dg(b) \circ Df(a)h = A$$

with $A = Dg(b)s(h) + r(y - b)$.

Let us show that $A = o(\| h \|)$. First $\| Dg(b)s(h) \| \leqslant \| Dg(b) \| \; \| s(h) \|$ shows that $Dg(b)s(h) = o(\| h \|)$. Next, given any $\varepsilon > 0$, we can choose $\| h \|$ so small that $\| y - b \| = \| Df(a)h + s(h) \| \leqslant (\| Df(a) \| + \varepsilon) \| h \|$. Thus $\| y - b \| / \| h \| \leqslant \| Df(a) \| + \varepsilon$. On the other hand, the continuity of f at a implies that $y \to b$ if $h \to 0$. Hence $r(y - b) = o(\| h \|)$, and so $A = o(\| h \|)$.

Tangent map

Suppose that f: $U \subset E \to F$ is differentiable on U. The tangent map Tf: $U \times E \to F \times F$ is defined by $Tf(a, h) = (f(a), Df(a)h)$.

Here is its *geometrical interpretation*. Take a differentiable curve in E, with origin a, that is, a differentiable map c: $\mathbf{R} \to E$ such that $c(0) = a$. We may think of $c(t)$ as the position at time t of a point moving in E. Let $c'(0) = h$ be its velocity vector at the origin. By the preceding theorem the curve $f \circ c : \mathbf{R} \to F$, the image of the curve c under the map f, is differentiable and its tangent vector at the origin is:

$$(f\circ c)'(0)=D(f\circ c)(0).1=Df(c(0))\circ Dc(0).1=Df(a)c'(0)=Df(a)h.$$

Thus to the pair consisting of a point on a curve and the velocity vector at this point, Tf makes correspond the pair consisting of the image of the point and the velocity vector of this image.

It is easy to see that the last theorem may be written as $T(g\circ f)=Tg\circ Tf$; T is thus a covariant functor. This is not the case for D, as $D(g\circ f)=(Dg\circ f)\circ Df$ and not $Dg\circ Df$.

1.2.7 Mapping into a direct sum

Let $F=F_1\oplus\ldots\oplus F_n$ be a direct sum of normed v.s. (§A.1.4). Denote by $p_r:F\to F_r$ the projection $(x_1,\ldots,x_n)\rightsquigarrow x_r$ and by $i_r:F_r\to F$ the injection which makes correspond to x_r the vector all of whose components are zero except for the rth, which equals x_r.

A map f of an open subset U of a normed v.s. E to F may be written as $x\rightsquigarrow(f_1(x),\ldots,f_n(x))$, where $f_r(x)$ is the rth component of $f(x)$. Thus:

$$f_r=p_r\circ f,\quad f=\sum_{r=1}^{n} i_r\circ f_r \tag{1.2}$$

Theorem 1.1

The map f is differentiable at $a\in U$ if, and only if, $f_1,\ldots,f_n$ are too. Under these conditions $Df(a)=(Df_1(a),\ldots,D_nf(a))$; that is:

$$Df(a)=\sum_{r=1}^{n} i_r\circ Df_r(a)$$

It follows that f is C^1 on U if, and only if, $f_1,\ldots,f_n$ are too.

Proof. This is an immediate consequence of the chain rule and equation (1.2). □

Corollary. If $f:x\in\mathbf{R}^n\rightsquigarrow(f_1(x),\ldots,f_m(x))\in\mathbf{R}^m$ is differentiable at a, the rth row of its Jacobian matrix (§1.1) at a is the Jacobian matrix at a of $f_r:\mathbf{R}^n\to\mathbf{R}$. □

1.2.8 Leibnitz' formula

Let E, F_1, F_2 and G be normed v.s., let U be an open subset of E, let $f: U\to F_1$ and $g: U\to F_2$ be maps which are differentiable at $a\in U$ (resp. C^1 on U), and let $b: F_1\times F_2\to G$ be a continuous bilinear map. Define $p=b(f,g): U\to G$ by $x\ \to b(f(x),g(x))$. Then p is differentiable at a (resp. C^1 on U) and $Dp(a)h=b(Df(a)h,g(a))+b(f(a),Dg(a)h)$ for all $h\in E$.

Proof. Since p is the composite map $x \rightsquigarrow (f(x), g(x)) \rightsquigarrow b(f(x), g(x))$, the theorem follows from the chain rule and from §§1.2.3 and 1.2.7:

$$\begin{aligned} Dp(a)h &= Db(f(a), g(a)) \circ D(f, g)(a)h \\ &= Db(f(a), g(a)) \circ (Df(a)h, Dg(a)h) \\ &= b(Df(a)h, g(a)) + b(f(a), Dg(a)h) \end{aligned}$$ □

Special cases

If $E = K$, Remark (b) in §1.1 shows that $Df(a)h$ may be identified with the product $hf'(a)$ of the scalar $h \in K$ and the derivative vector $f'(a)$. Similarly $Dg(a)h = h \cdot g'(a)$. Leibnitz' formula then becomes, on setting $h = 1$:

$$p'(a) = b(f'(a), g(a)) + b(f(a), g'(a))$$

If $F_1 = F_2 = G = K$ and $b(u, v) = uv$, we thus have the formula for the derivative of a product fg of scalar-valued functions: $(fg)' = f'g + fg'$.

If $F_1 = F_2 = G$ is the space $\mathbf{R}^3$, oriented and endowed with its usual scalar product, and if b is the vector product, we obtain $(f \wedge g)'(a) = f'(a) \wedge g(a) + f(a) \wedge g'(a)$. An analogous formula may be obtained for the scalar product.

1.2.9 Map defined on a direct sum

Partial derivatives

Let E be the direct sum $E_1 \oplus \ldots \oplus E_n$ of normed v.s., and let U be an open subset of E. A map f of U to a normed v.s. F may then be written as $(x_1, \ldots, x_n) \rightsquigarrow f(x_1, \ldots, x_n)$; it is a function of the n variables $x_1 \in E_1, \ldots, x_n \in E_n$.

Let $a = (a_1, \ldots, a_n) \in E$ be given. The map

$$x_r \in E_r \rightsquigarrow f(a_1, \ldots, a_{r-1}, x_r, a_{r+1}, \ldots, a_n) \in F \tag{1.3}$$

(the rth component a_r is replaced by x_r) is the composite map $f \circ I_r$, where $I_r : x_r \rightarrow (a_1, \ldots, a_{r-1}, x_r, a_{r+1}, \ldots, a_n)$ is clearly a continuous map of E_r to the direct sum $E_1 \oplus \ldots \oplus E_n$. Hence $f \circ I_r$ is defined on the open subset $I_r^{-1}(U)$ of E_r, which contains a_r.

If the mapping in equation (1.3), that is $f \circ I_r$, is differentiable at a_r, we call its derivative at a_r the *partial derivative* of f with respect to the rth variable at the point a, and denote it by $D_r f(a)$; it is an element of $\mathscr{L}(E_r; F)$.

If $E_1 = \ldots = E_n = K$, $D_r f(a)$ is called, more classically, the partial derivative with respect to the rth variable at the point a. This is the usual derivative $g'(a_r)$ of the vector-valued function of a scalar variable:

$$x \rightsquigarrow g(x) = f(a_1, \ldots, a_{r-1}, x, a_{r+1}, \ldots, a_n)$$

at the point $x = a_r$. It is also denoted by $(\partial f / \partial x_r)(a)$.

For example, if f: $\mathbf{R}^2 \rightarrow \mathbf{R}$ is defined by $f(x, y) = \sin(x^2 y)$, then $D_1 f(a, b)$

is the derivative at a of $x \rightsquigarrow \sin(x^2 b)$; that is, $2ab\cos(a^2 b)$. Similarly, $D_2 f(a, b)$ is the derivative at b of $y \rightsquigarrow \sin(a^2 y)$; that is, $a^2 \cos(a^2 b)$.

Let us now return to the general case and its notation.

Theorem 1.2

If f is differentiable at a, then each of its partial derivatives $D_r f(a)$, $r = 1, \ldots, n$, exists; and if $h = (h_1, \ldots, h_n) \in E$ we have:

$$Df(a)h = \sum_{r=1}^{n} D_r f(a) h_r.$$

Proof. If $i_r : E_r \to E$ is the injection $x_r \rightsquigarrow (0, \ldots, 0, x_r, 0, \ldots, 0)$ already considered in §1.2.7, we have $I_r(x_r) = a + i_r(x_r - a_r)$. Thus I_r is an affine map, and hence $DI_r(a_r) = i_r$ (§1.2.2). The first part now follows from the chain rule, and

$$D_r f(a) = D(f \circ I_r)(a_r) = Df(a) \circ i_r. \tag{1.4}$$

But, if $p_r : E \to E_r$ is the projection on the rth component already considered in §1.2.7, clearly $f = f \circ \sum_{r=1}^{n} i_r \circ p_r$. The chain rule and equation (1.4) give:

$$Df(a) = \sum_{r=1}^{n} Df(a) \circ i_r \circ p_r = \sum_{r=1}^{n} D_r f(a) \circ p_r$$

and thus:

$$Df(a) \cdot h = \sum_{r=1}^{n} D_r f(a) h_r. \qquad \square$$

Remarks

(a) The existence of partial derivatives does not imply differentiability. For example, if we take $f: \mathbf{R}^2 \to \mathbf{R}$ to be defined by:

$$f(x, y) = \begin{cases} xy(x^2 + y^2)^{-1/2} & \text{if } x^2 + y^2 \neq 0 \\ 0 & \text{if } x = y = 0 \end{cases}$$

the function is zero for $x = 0$ (resp. $y = 0$), and so $D_1 f(0, 0) = D_2 f(0, 0) = 0$. If f were differentiable at $(0, 0)$, its derivatives at that point would be zero, by Theorem 1.2, and we would have $f(h_1, h_2) = o(\|h\|)$, where $\|h\| = (h_1^2 + h_2^2)^{1/2}$. However, if $h = (3t, 4t)$, we have $f(3t, 4t)/\|h\| = 12/25$, which does not tend to zero with t.

Furthermore, a function f may have partial derivatives at a point without being continuous there. Consider at $(0, 0)$ the function

$$f(x, y) = \begin{cases} xy(x^2 + y^2)^{-1} & \text{if } x^2 + y^2 \neq 0 \\ 0 & \text{if } x = y = 0 \end{cases}$$

The reason for this 'pathology' is that the existence of partial derivatives at $(0, 0)$ guarantees the existence of tangents at the origin to the curves $x \rightsquigarrow f(x, 0)$ and $y \rightsquigarrow f(0, y)$, which are the intersection of the surface with equation $z = f(x, y)$ with the planes $y = 0$ and $x = 0$, but it says nothing about the behaviour of the surface outside these planes.

(b) If f is C^1 on U, so are its partial derivatives, since equation (1.4) shows that $D_r f(a) = Df(a) \circ i_r$ depends continuously on a. In Chapter 2 we shall prove a converse to this.

1.2.10 Calculation of the Jacobian matrix

If f: $x = (x_1, \ldots, x_n) \in \mathbf{R}^n \rightsquigarrow (f_1(x), \ldots, f_m(x)) \in \mathbf{R}^m$ is differentiable on an open set containing a, the element at the ith row and rth column of its Jacobian matrix at a is $D_r f_i(a)$.

Theorem 1.3

By §1.2.7, the ith row of the Jacobian matrix at a is the Jacobian matrix at a of $f_i: \mathbf{R}^n \to \mathbf{R}$. By §1.2.9, the Jacobian matrix of f_i at a is $(D_1 f_i(a), \ldots, D_n f_i(a))$.

We can give an alternative proof. If (e_i) is the canonical basis of $\mathbf{R}^n$, consider the differentiable curve c: $t \in \mathbf{R} \rightsquigarrow a + te_r \in R^n$ and the composite map $(f \circ c)(t) = f(a + te_r) = (f_1(a + te_r), \ldots)$. On applying the chain rule and Remark (b) in §1.1, and then setting $t = 0$, we have:

$$Df(a)e_r = \frac{\mathrm{d}}{\mathrm{d}t} f(a + te_r)\,|_{t=0} = (D_r f_1(a), \ldots, D_r f_m(a)). \qquad \square$$

The following theorem demonstrates an important practical application, the changing of variables.

Theorem 1.4

Let $g_1, \ldots, g_m: \mathbf{R}^n \to \mathbf{R}$ be differentiable at a, and let f: $\mathbf{R}^m \to \mathbf{R}$ be differentiable at $b = (g_1(a), \ldots, g_m(a))$. Define F: $\mathbf{R}^n \to \mathbf{R}$ by $F(x) = f(g_1(x), \ldots, g_m(x))$. Then:

$$D_i F(a) = \sum_{r=1}^{n} D_r f(b) D_i g_r(a).$$

Proof. If g: $\mathbf{R}^n \to \mathbf{R}^m$ is defined by $g(x) = (g_1(x), \ldots, g_m(x))$, then $F = f \circ g$. The chain rule gives $DF(a) = Df(g(a)) \circ Dg(a) = Df(b) \circ Dg(a)$. It is now enough to note that the Jacobian matrix of F at a is equal to the product of the Jacobian matrices of f at b and of g at a, and that these matrices are given by Theorem 1.3. $\square$

Example

We calculate the partial derivatives of the map $F: \mathbf{R}^2 \to \mathbf{R}$ defined by $F(x, y) = f(h(x), s(y))$, where f, h and s are differentiable. We reduce the problem to the preceding theorem by setting $g_1(x, y) = h(x)$ and $g_2(x, y) = s(y)$. Thus:

$$F(x, y) = f(g_1(x, y), g_2(x, y))$$

and

$$D_1F(x, y) = D_1f(u)D_1g_1(x, y) + D_2f(u)D_1g_2(x, y)$$
$$D_2F(x, y) = D_1f(u)D_2g_1(x, y) + D_2f(u)D_2g_2(x, y)$$

where for shortness we have put $u = (g_1(x, y), g_2(x, y))$.

But $D_1g_1(x, y) = h'(x)$, $D_1g_2(x, y) = 0$, $D_2g_1(x, y) = 0$ and $D_2g_2(x, y) = s'(y)$. Thus $D_1F(x, y) = D_1f(u)h'(x)$ and $D_2F(x, y) = D_2f(u)s'(y)$.

2

Mean-value theorems

This chapter is devoted to the generalization of the classical mean-value theorem for scalar-valued functions. Among other applications, the fundamental theorem of integral calculus is proved.

PRELIMINARIES

Let f be a real-valued function defined and continuous on an interval $[a, b]$ and differentiable on $]a, b[$. The classical mean-value theorem says that there exists $c \in]a, b[$ such that $f(b) - f(a) = (b-a)f'(c)$. In geometrical terms, this means that there is a point $(c, f(c))$ of the graph of f where the tangent is parallel to the line joining the end-points $(a, f(a))$ and $(b, f(b))$ of this graph.

Now if, under the same hypothesis, f is a function with values in $\mathbf{R}^n$, $n \geqslant 2$, the above geometrical property may not hold: think of the graph of a 'corkscrew', where the tangent at a point cannot be parallel to a line joining two points of the graph. An example is f: $[0, \pi/2] \to \mathbf{R}^2$ defined by $f(t) = (\cos t, \sin t)$.

Let us give another interpretation of the mean-value theorem. If f: $[a, b] \to F$ is a continuous map to a normed v.s. F, differentiable on $]a, b[$, we may think of $f(t)$ (resp. $f'(t)$) as the position at time t of a point moving in F. The speed is $\|f'(t)\|$. Suppose that a second point starts from $f(a)$ at the same time as the first, that it describes a straight line and that its speed $g'(t)$ is, at each instant, at least equal to that of the first point. It is intuitively clear that the second point will recede from the point of departure $f(a)$ more quickly than the first. Let us make this rigorous.

2.1 MEAN-VALUE THEOREMS

2.1.1 General results

Theorem 2.1

Let a and b be real numbers such that $a < b$, let F be a normed v.s., and let f: $[a, b] \to F$ and g: $[a, b] \to R$ be maps which are continuous on $[a, b]$ and differentiable on $]a, b[$. Suppose that $\|(f)'(t)\| \leqslant g'(t)$ for $a < t < b$. Then $\|f(b) - f(a)\| \leqslant g(b) - g(a)$.

Proof. We shall show that if u and v are real numbers such that $a < u < v < b$, then $\|f(v) - f(u)\| \leqslant g(v) - g(u)$. Since f and g are continuous at a and b, the theorem will follow by letting u approach a and v approach b.

Suppose the result is false. Then $\|f(v) - f(u)\| - [g(v) - g(u)] = M > 0$. Divide $[a_0 = u, b_0 = v]$ at its mid-point $m = (u + v)/2$. The triangle inequality shows that on one of the intervals $[u, m]$ and $[m, v]$ we have:

$$\|f(v) - f(m)\| - [g(v) - g(m)] \geqslant M/2$$

or

$$\|f(m) - f(u)\| - [g(m) - g(u)] \geqslant M/2.$$

Denote by $[a_1, b_1]$ one of these intervals on which the inequality holds. Repeat the procedure by dividing $[a_1, b_1]$ at its mid-point, etc. We obtain a sequence of intervals $[a_n, b_n]$ such that $a_0 \leqslant \ldots \leqslant a_n \leqslant b_n \leqslant \ldots \leqslant b_0$, $|b_n - a_n| = (b - a)/2^n$, and

$$\|f(b_n) - f(a_n)\| - [g(b_n) - g(a_n)] \geqslant M/2^n.$$

It follows that a_n and b_n converge to the same point $w \in]a, b[$ and that

$$\begin{aligned} M/2^n &\leqslant \|f(b_n) - f(w)\| + \|f(w) - f(a_n)\| - [g(b_n) - g(w)] - \\ &\quad [g(w) - g(a_n)] \\ &\leqslant \|Df(w)(b_n - w)\| + o(b_n - w) + \|Df(w)(w - a_n)\| + o(w - a_n) \\ &\quad - [g'(w)(b_n - w) + o(b_n - w)] - [g'(w)(w - a_n) + o(w - a_n)] \\ &\leqslant \|Df(w)\| (|b_n - w| + |w - a_n|) - g'(w)(|b_n - w| + |w - a_n|) \\ &\quad + o(b_n - w) + o(w - a_n) \\ &= [\|Df(w)\| - g'(w)] (b_n - a_n) + o(b_n - a_n). \end{aligned}$$

Divide both sides by $b_n - a_n = (b - a)/2^n$ and let n tend to ∞. We obtain the contradiction $M/(b - a) \leqslant \|Df(w)\| - g'(w)$. □

The intuition which led to Theorem 2.1 makes us guess that pauses, followed by departures in new directions, are permitted in the movement of the point $f(t)$. This follows from a stronger theorem, the proof of which can be found in texts by Cartan and Dieudonné (see the Bibliography).

Theorem 2.2

Let F be a normed v.s., and let f: $[a, b] \to F$ and g: $[a, b] \to \mathbf{R}$ be continuous. Suppose that for all $t \in [a, b]$, except perhaps for a countable set of points, f and g are differentiable and $\|f'(t)\| \leqslant g'(t)$. Then $\|f(b) - f(a)\| \leqslant g(b) - g(a)$. □

Corollary 2.1

Let f: $[a,b] \to F$ be a continuous map into a normed v.s. such that $f'(t)$ exists for all $t \in]a,b[$. Suppose there is a constant k such that $\|f'(t)\| \leqslant k$ for all $t \in]a,b[$. Then $\|f(b)-f(a)\| \leqslant k(b-a)$.

Proof. This can be shown by taking $g(t)=kt$ in Theorem 2.2. □

We shall now assume that f is defined on an open subset U of a normed v.s. E, which need no longer be **R**.

Corollary 2.2

If f: $U \to F$ is differentiable on U, and if the line segment $[a,b] = \{(1-t)a+tb\colon 0 \leqslant t \leqslant 1\}$, with end-points $a, b \in U$, is contained in U, then:

$$\|f(b)-f(a)\| \leqslant \sup_{0 \leqslant t \leqslant 1} \|Df[(1-t)a+tb]\| \, \|b-a\|$$

Proof. The chain rule, applied to

$$t \in [0,1] \rightsquigarrow h(t) = f[(1-t)a+tb]$$

gives $h'(t) = Df[(1-t)a+tb](b-a)$. Thus $\|h'(t)\| \leqslant \sup_{0 \leqslant t \leqslant 1} \|Df[\ ]\|$ $\|b-a\|$, and it is enough to apply Corollary 2.1, replacing a by 0, b by 1, f by h and k by $\sup \|Df[\ ]\| \, \|(b-a\|$. □

2.1.2 Convexity

We say that a subset U of a v.s. is convex if, for all a, $b \in U$, the line segment joining these points lies in U.

Theorem 2.3

Let U be a convex open subset of a normed v.s. E, and let f be a differentiable map of U to a normed v.s. F. If there exists $k \geqslant 0$ such that $\|Df(u)\| \leqslant k$ for all $u \in U$, then:

$$\|f(b)-f(a)\| \leqslant k\|b-a\| \quad \text{for all} \quad a, b \in U.$$

Proof. This is an immediate consequence of Corollary 2.2. □

Corollary 2.3

Let U be a convex open subset of a normed v.s. E, and let f be a differentiable map of U to a normed v.s. F. Then for all $a, b, c \in U$, we have:

$$\|f(b)-f(a)-Df(c)(b-a)\| \leqslant \sup_{u \in U} \|Df(u)-Df(c)\| \, \|b-a\|.$$

Proof. Apply Corollary 2.2 to $u \mapsto f(u) - Df(c)u$, the derivative of which is $Df(u) - Df(c)$. □

Let us now move on to applications of the preceding results.

2.2 CONVERSE OF §1.2.2

Theorem 2.4

Let U be a connected open subset of a normed v.s. E, and let f be a differentiable map of U to a normed v.s. F. If $Df(u) = 0$ for all u in U, then f is constant.

Proof. Fix a in U and denote by B the set of all points $b \in U$ such that $f(b) = f(a)$. Since f is continuous, B is closed in U. On the other hand, every x in B is the centre of an open ball contained in U and of positive radius. This ball is convex. By Theorem 2.3, $f(y) = f(x) = f(a)$ for all y in this ball. Hence B is also open in U. As B is non-empty and U is connected, $B = U$. □

2.3 NECESSARY AND SUFFICIENT CONDITION FOR A MAP TO BE OF CLASS C^1

Let us recall §1.2.9. Let U be an open subset of the direct sum $E = E_1 \oplus \ldots \oplus E_n$ of normed v.s., and let f be a map of U to a normed v.s. F. We have seen that, if f is C^1 in U, then the partial derivatives $D_k f\colon U \to \mathscr{L}(E_k; F)$ are C^1 in U. We shall prove the converse.

Theorem 2.5

With the above notation, suppose that the partial derivatives exist at each point $x = (x_1, \ldots, x_n)$ of U and that the maps $D_k f\colon U \to \mathscr{L}(E_k; F)$ are continuous at $a \in U$. Then f is C^1 at a.

Proof. By the formula $Df(a)h = \sum_{k=1}^{n} D_k f(a)h_k$ of §1.2.9, it is enough to show that $Df(a)$ exists. Define

$$g(x) = f(x) - f(a) - \sum_{k=1}^{n} D_k f(a)(x_k - a_k).$$

Then $D_k g(x) = D_k f(x) - D_k f(a)$.

Since the partial derivatives are continuous at a, given any $\varepsilon > 0$, there exists $r > 0$ such that $\| D_k g(x) \| \leqslant \varepsilon$, $k = 1, \ldots, n$, if x is in the open ball with centre a and radius r. It follows from Theorem 2.3 that:

$$\| g(x) \| \leqslant \sum_{k=1}^{n} \| g(x_1, \ldots, x_k, a_{k+1}, \ldots, a_n) - g(x_1, \ldots, x_{k-1}, a_k, \ldots, a_n) \|$$

$$\leqslant \sum_{k=1}^{n} \varepsilon \| x_k - a_k \| \leqslant n\varepsilon \| x - a \|.$$

As ε is arbitrary:

$$f(x) = f(a) + \sum_{k=1}^{n} D_k f(a)(x_k - a_k) + o(\| x - a \|). \qquad \square$$

2.4 A CRITERION FOR UNIFORM CONVERGENCE

Theorem 2.6

Let U be a connected open subset of a normed v.s. E, and let $f_n : U \to F$ be a sequence of differentiable maps into a Banach space F. Suppose that: (1) there is a point a of U such that the sequence $f_n(a)$ converges; (2) the sequence $Df_n : U \to \mathscr{L}(E; F)$ converges uniformly on each bounded subset of U to a map g: $U \to \mathscr{L}(E; F)$. Then for each $x \in U$, the sequence $f_n(x)$ converges to a limit, denoted by $f(x)$. This convergence is uniform on each bounded convex subset of U. Finally, $f = \lim_{n\to\infty} f_n$ is differentiable and $Df = g$.

Proof. Let B be a ball, with centre a and radius $r > 0$, contained in U. By Corollary 2.2, for all $x \in B$ we have:

$$\| f_p(x) - f_q(x) - [f_p(a) - f_q(a)] \| \leqslant \sup_{u \in B} \| Df_p(u) - Df_q(u) \| \; \| x - a \|. \tag{2.1}$$

Since the sequence $f_n(a)$ converges and the sequence $Df_n(u)$ is uniformly convergent on B, the sequence $f_n(x)$ is a Cauchy sequence in F. As F is *complete*, $\lim_{n\to\infty} f_n(x) = f(x)$ exists. This reasoning also shows that if f converges at one point of an open ball in U, it converges uniformly on that ball. The set of points u of U at which $f_n(u)$ converges is thus open and closed in U. Since it contains a and U is connected, it coincides with U. In other words, $\lim_{n\to\infty} f_n(u) = f(u)$ exists for all $u \in U$.

Now denote by B a bounded *convex* subset of U, of diameter $d = \sup_{x,y \in B} \| x - y \|$. By Corollary 2.2, inequality (2.1) still holds if $a, x \in B$, and thus:

$$\| f_p(x) - f_q(x) \| \leqslant \| f_p(a) - f_q(a) \| + \sup_B \| \; \| d$$

This shows that f_n converges uniformly on B.

Consider inequality (2.1) again and let p tend to $+\infty$. Since $\lim Df_n = g$, we have:

$$\| f(x) - f(a) - [f_q(x) - f_q(a)] \| \leqslant \sup_{u \in B} \| g(u) - Df_q(u) \| \; \| x - a \|.$$

Since the convergence of Df_q is uniform on B, given any $\varepsilon > 0$, there exists $N > 0$ such that $q > N$ implies that $\sup_B \| g(u) - Df_q(u) \| < \varepsilon$. On the other hand, given any such q, there exists $r' \leqslant r$ such that $\| x-a \| \leqslant r'$ implies that:

$$\| f_q(x) - f_q(a) - Df_q(a)(x-a) \| \leqslant \varepsilon \| x-a \|.$$

The triangle inequality and the preceding inequalities now show that:

$$\| f(x) - f(a) - g(a)(x-a) \| \leqslant 3\varepsilon \| x-a \|.$$

Thus f is differentiable at a and $Df(a) = g(a)$. □

We deduce from this the following theorem.

Theorem 2.7

Let U be a connected open subset of a normed v.s. E, and let f_n be a sequence of differentiable maps of U to a Banach space F. Suppose that: (1) there is a point a of U such that the series $\sum f_n(a)$ converges; (2) the series $\sum Df_n$ converges uniformly on every bounded subset of U to a map $S: U \to \mathscr{L}(E; F)$. Then for each x in U the series $\sum f_n(x)$ converges to a limit, denoted by $h(x)$. This convergence is uniform on each bounded convex subset of U. Lastly, h is differentiable and $Dh = S$. □

2.5 SARD'S THEOREM

Definition 2.1

Let $f: U \to \mathbf{R}^n$ be a differentiable map of an open subset U of $\mathbf{R}^p$ to $\mathbf{R}^n$. We say that $a \in U$ is a *critical point* of f if the rank of the linear map $Df(a)$ is less than n.

Definition 2.2

Give $\mathbf{R}^n$ its usual scalar product. This scalar product defines the usual norm and distance. A displacement of $\mathbf{R}^n$ is an affine map of $\mathbf{R}^n$ into $\mathbf{R}^n$ which preserves this distance.

Let $a_1, \ldots, a_n \in \mathbf{R}$ and $h_1 > 0, \ldots, h_n > 0$ be given. Put:

$$P = [a_1, a_1 + h_1] \times \ldots \times [a_n, a_n + h_n]$$

We shall understand by a box in $\mathbf{R}^n$ any image of P under a displacement. The volume of a box is, by definition, $h_1 \ldots h_n$.

Definition 2.3

A subset E of $\mathbf{R}^n$ is of measure zero if, given any $\varepsilon > 0$, there is a covering of E by boxes the sum of whose volumes is less than ε.

It is left as an exercise for the reader to show that the union of a countable family E_n, $n = 1, 2, \ldots$, of sets of measure zero is itself of measure zero (hint: cover E_n by boxes the sum of whose volumes is less than $\varepsilon/2^n$).

Theorem 2.8

Let $f: U \to \mathbf{R}^n$ be a C^1 map of an open subset U of $\mathbf{R}^p$ to $\mathbf{R}^n$. Then the image $f(C)$ of the set C of critical points of f is a set of measure zero in $\mathbf{R}^n$.

Proof. The proof is difficult if $p > n$ (see J. Milnor). We shall limit ourselves to the case $p = n$.

Step 1: Cover $\mathbf{R}^n$ by cubes with sides of length 1 and with vertices with integer coordinates; $\mathbf{R}^n$ is a countable union of such cubes. By Definition 2.3, it is enough to show that the image under f of the set of critical points contained in one of these cubes is of measure zero. Application of a translation enables us to assume that this cube is the unit cube $I = [0, 1]^n$. It remains to prove that $f(C \cap I)$ is of measure zero.

Step 2: Let $x \in C \cap I$. Since rank $Df(x) < n$, $Df(x)\mathbf{R}^n$ is a subspace of dimension less than or equal to $n - 1$. Hence there exists a hyperplane P, passing through $f(x)$ and containing all vectors of the form $f(x) + Df(x)z$, $z \in \mathbf{R}^n$. Fix $\varepsilon > 0$ and let y belong to the open ball $B(x, \varepsilon)$ with centre x and radius ε. By Corollary 2.3 we have:

$$f(y) = f(x) + Df(x)(y - x) + b(\varepsilon) \| y - x \|$$

where $b(\varepsilon) \leqslant \sup_{u \in B} \| Df(x) - Df(u) \|$ tends to zero with ε, since Df is continuous and thus *uniformly continuous on the compact set I.* As $f(x) + Df(x)(y - x) \in P$, we see that the distance of $f(y)$ from the hyperplane P is dominated by $\varepsilon b(\varepsilon)$. Thus $f[B(x, \varepsilon)]$ lies between two hyperplanes parallel to P and whose distance to P is $\varepsilon b(\varepsilon)$.

On the other hand, Corollary 2.2 shows that $\| f(y) - f(x) \| \leqslant \sup_I \| Df(u) \| \; \| y - x \|$. It follows that $f(y)$ lies in the ball with centre $f(y)$ and radius $a\varepsilon$, where $a = \sup_I \| Df(u) \|$.

To sum up, $f[B(x, \varepsilon)]$ lies in a right cylinder, whose base is the intersection of P with the ball of centre $f(x)$ and radius $a\varepsilon$, and whose height is $2\varepsilon b(\varepsilon)$. It is clear that $f[B(x, \varepsilon)]$ is contained in a box in $\mathbf{R}^n$, whose sides parallel to P are of length $(\sqrt{n})ab$ and whose side perpendicular to P is of length $2\varepsilon b(\varepsilon)$. The volume of such a box is $2n^{(n-1)/2} a^{n-1} \varepsilon^n b(\varepsilon)$.

Step 3: Divide each of the n sides of the unit cube into k equal parts. We obtain k^n cubes of side $1/k$. Each of them is contained in a ball of radius $\varepsilon = \sqrt{n}/k$ (Pythagoras' theorem!). Some of them will intersect the set C of critical points. Their image under the map f is thus contained, by step 2, in a box of volume:

$$2n^{(n-1)/2} a^{n-1} b(\sqrt{n}/k) k^{-n}.$$

It follows that $f(C)$ is contained in the union of at most k^n boxes, the sum of whose volumes, $2n^{(n-1)/2}a^{n-1}b(\sqrt{n}/k)$, tends to zero as $k \to +\infty$. □

2.6 INTEGRATION OF REGULATED FUNCTIONS, AND THE FUNDAMENTAL THEOREM OF INTEGRAL CALCULUS

Finally in this chapter we shall look at an extremely important application of mean-value theorems. All the spaces in this section are *Banach spaces*.

2.6.1 Step functions

Let F be a Banach space, and let $[a, b]$, $a < b$, be a closed bounded interval in $\mathbf{R}$. Partition $[a, b]$ into n intervals by points $a = a_0 < a_1 < \dots < a_n = b$; let $C_1, \dots, C_n$ be n given vectors in F. The function f: $[a, b] \to F$ defined by $f(x) = C_1$ for $a_0 \leqslant x < a_1$, $f(x) = C_2$ for $a_1 \leqslant x < a_2, \dots$, and $f(x) = C_n$ for $a_{n-1} \leqslant x \leqslant a_n$, is called a step function. The set of all such functions is denoted by $([a, b], F)$.

The sum of two step functions, and the product of a scalar and a step function, are also step functions. It follows that $([a, b], F)$ is a vector space. We shall prove the first assertion. Let f, $g \in ([a, b], F)$; suppose that $f(x) = C_i$ for $a_i \leqslant x < a_{i+1}$, $0 \leqslant i \leqslant n$, and $g(x) = C_j'$ for $b_j \leqslant x < b_{j+1}$, $0 \leqslant j \leqslant m$. Form a partition of $[a, b]$ by inserting between a and b the points a_i, b_j. In any interval of this partitition f and g are constant, and thus so is $f + g$.

If $f \in ([a, b], F)$ we put $\| f \|_{C^0} = \sup_{[a,b]} \| f(x) \|$. It is clear that $\| \; \|_{C^0}$ is a norm: it is the norm of uniform convergence on $[a, b]$.

2.6.2 Regulated functions

The space $\mathscr{R} = \mathscr{R}([a, b], F)$ of regulated functions is the completion of the normed space $([a, b], F)$. A regulated function is thus the uniform limit of a sequence of step functions. For example, a continuous function h: $[a, b] \to F$ is regulated. For, since h is continuous on the compact set $[a, b]$, it is uniformly continuous. Hence given any integer $n > 0$ there exists $p > 0$ such that $\| h(x) - h(y) \| \leqslant 1/n$ if $x, y \in [a, b]$ and $|x - y| < p$. Choose a partition $a = a_0 < a_1 \dots < a_m = b$ such that $|a_{i+1} - a_i| < p$ for $i = 0, 1, \dots, m-1$, and define f_n by $f_n(x) = h(a_1)$ for $a_0 \leqslant x < a_1$, $f_n(x) = h(a_2)$ for $a_1 \leqslant x < a_2, \dots, f_n(x) = h(a_m)$ for $a_{m-1} \leqslant x \leqslant a_m$.

Clearly $\| h - f_n \|_{C^0} \leqslant 1/n$, and the sequence of step functions f_n converges uniformly to h on $[a, b]$.

2.6.3 Integral of a step function

If $f(x) = C_1$ for $a_0 \leqslant x \leqslant a_1, \dots,$ and $f(x) = C_n$ for $a_{n-1} \leqslant x \leqslant a_n$, the integral of f from a to b is defined to be:

$$I(f)=\int_a^b f(x)\,\mathrm{d}x=\sum_{i=0}^{n-1}(a_{i+1}-a_i)C_{i+1}.$$

It is easy to verify that I: $([a,b],F)\to F$ is linear. Moreover, the triangle inequality in F shows that $\|I(f)\|\leqslant(b-a)\|f\|_{C^0}$. The linear map I is thus continuous.

Finally, if u lies between a and b, we see that:

$$\int_a^b f(x)\,\mathrm{d}x=\int_a^u f(x)\,\mathrm{d}x+\int_u^b f(x)\,\mathrm{d}x$$

If we write $\int_u^v=-\int_v^u$ when $v<u$, the relation (called Chasles' relation) follows:

$$\int_u^v f(x)\,\mathrm{d}x=\int_u^w f(x)\,\mathrm{d}x+\int_w^v f(x)\,\mathrm{d}x\quad\forall u,v,w\in[a,b].$$

2.6.4 Integral of a regulated function

If $f\in\mathscr{R}$ is the uniform limit of a sequence of step functions f_n, the sequence $I(f_n)$ converges. For since I is continuous and linear, $I(f_n)$ is a Cauchy sequence in F. *Since F is complete*, $I(f_n)$ converges (this is why all the spaces in this section are Banach spaces). In fact, $\lim_{n\to\infty} I(f_n)$ does not depend on the sequence used to approximate f uniformly: if g_p is another sequence which converges to f, then:

$$\|f_n-g_p\|_{C^0}\leqslant\|f_n-f\|_{C^0}+\|f-g_p\|_{C^0}$$

which tends to zero if $n,\ p\to+\infty$; thus $\|I(f_n)-I(g_p)\|\leqslant(b-a)\|f_n-g_p\|_{C^0}$ tends to zero as $n,\ p\to+\infty$. We may therefore define the integral of $f\in\mathscr{R}$ from a to b by:

$$I(f)=\int_a^b f(x)\,\mathrm{d}x=\lim_{n\to\infty} I(f_n).$$

The following properties are derived from §2.6.3.

(a) *I: $\mathscr{R}\to F$ is linear.* For if f_n (resp. g_n) is a sequence of step functions which converges uniformly to the regulated function f (resp. g), then f_n+g_n converges uniformly to $f+g$, and:

$$\begin{aligned}I(f+g)&=\lim I(f_n+g_n)=\lim[I(f_n)+I(g_n)]=\lim I(f_n)+\lim I(g_n)\\&=I(f)+I(g).\end{aligned}$$

Similarly, $I(kf)=kI(f)$ if k is a scalar.

(b) $\|I(f)\|\leqslant(b-a)\|f\|_{C^0}$; this follows easily from the analogous inequality for step functions and from the continuity of $X\mapsto\|X\|_F$. In particular, I is a continuous linear map.

(c) The Chasles relation holds for regulated functions. (Routine verification.)

(d) Lastly, if $L: F \to K$ (scalar field) is a continuous linear map, and if $f \in \mathcal{R}$, then $L \circ f$ is a regulated function and $L[I(f)] = I[L \circ f]$. For if f_n is a sequence of step functions which converges uniformly to f, then $L(f_n)$ is regulated and converges uniformly to $L(f)$.

2.6.5 Primitive of a regulated function

Let g be a function defined on an interval (a, b) (not necessarily bounded), with values in F and regulated on every closed bounded interval in (a, b).

To begin with it is convenient to write, just as in §2.6.3, $\int_u^v g = -\int_v^u g$ if $u, v \in (a, b)$ and $v < u$. Also, $\int_u^u g = 0$.

Let x_0 be a fixed number (arbitrary) in (a, b). The function G defined at $x \in (a, b)$ by $G(x) = \int_{x_0}^x g(t)\, dt$ is called a primitive of g.

Notice first that while there are infinitely many primitives (G depends on x_0), two primitives G_1 and G_2 of g differ only by a constant. For by Chasles' relation:

$$G_2(x) - G_1(x) = \int_{x_2}^{x} g(t)\, dt - \int_{x_1}^{x} g(t)\, dt = \int_{x_2}^{x_1} g(t)\, dt.$$

On the other hand, if g is continuous we have the following theorem.

Theorem 2.9

A primitive G of a continuous function g is of class C^1 and $G' = g$.

Proof. By Chasles' relation and §2.6.4, we have:

$$G(x+h) - G(x) - g(x)h = \int_x^{x+h} [g(t) - g(x)]\, dt$$

and the theorem follows from

$$\left\| \int_x^{x+h} [g(t) - g(x)]\, dt \right\| / \| h \| \leqslant \sup_{[x, x+h]} \| g(t) - g(x) \|$$

which tends to zero as $h \to 0$, since g is continuous. □

2.6.6 Fundamental theorem of integral calculus

Theorem 2.10

Let E and F be Banach spaces, let U be an open subset of E, and let $f: U \to F$ be a C^1 map. If $x + ty \in U$ for all $t \in [0, 1]$, then $f(x+y) = f(x) + \int_0^1 Df(x+ty)y\, dt$.

Proof. Set $g(t) = f(x+ty)$ for $0 \leqslant t \leqslant 1$. If $0 < t < 1$, the chain rule implies that $g'(t) = Df(x+ty)y$. Define $h(t) = f(x) + \int_0^t Df(x+sy)y\, ds$ for $0 \leqslant t \leqslant 1$.

By Theorem 2.9, $h'(t) = Df(x + ty)y$ for $0 < t < 1$. Hence $g'(t) - h'(t) = 0$ for $0 < t < 1$ and Theorem 2.4 implies that $g(t) - h(t)$ is constant for $0 < t < 1$. Since g and h are continuous, this relation also holds for $0 \leqslant t \leqslant 1$. But $g(0) = h(0) = f(x)$; and thus $g(1) = h(1)$, which proves the theorem. □

3

The concept of diffeomorphism, and the solution of equations

This chapter is devoted to solutions of the equation $f(x) = 0$. The existence of solutions is obtained by means of the convergence of successive approximations, the setting being that of complete spaces. This is why all the spaces in this chapter are Banach spaces.

PRELIMINARIES

Let $f\colon \mathbf{R} \to \mathbf{R}$ be a continuously differentiable map such that $f'(a) \neq 0$. Then there is an open interval I containing a, in which f' has constant sign (positive, say). Thus f is increasing on I and is a bijection of I on to the open interval $J = f(I)$.

If y is 'close enough' to $f(a)$, that is, if $y \in J$, the equation $f(x) = y$ therefore has a solution $x = f^{-1}(y)$. On the other hand we known that $f^{-1}: J \to I$ is continuously differentiable and that $(f^{-1})'(y) = [f'(x)]^{-1}$. The solution $x = f^{-1}(y)$ is thus a C^1 function of y.

It is this classical result that we are going to generalize.

3.1 DIFFEOMORPHISMS

Definition 3.1

Let U be an open subset of a normed v.s. E, and let V be an open subset of a normed v.s. F. We say that a map $f\colon U \to V$ is a diffeomorphism of U on V if f is a bijection, and if f and the inverse map f^{-1} are of class C^1.

Examples

(a) $x \mapsto \tan x$ is a diffeomorphism of $]-\pi/2, \pi/2[$ on $\mathbf{R}$.
(b) $z \mapsto z^2$ is a diffeomorphism of the half-plane $\{z \in \mathbf{C}\colon \operatorname{Im} z > 0\}$ on $\mathbf{C}\backslash\mathbf{R}^+$.

Remarks

(a) A diffeomorphism $f\colon U \to V$ is a homeomorphism, for f and f^{-1} are differentiable and thus continuous. But a homeomorphism of class C^1 is not necessarily a diffeomorphism, for f^{-1} may not be differentiable. Thus $x \rightsquigarrow x^3$ is a C^1 homeomorphism of $\mathbf{R}$ to $\mathbf{R}$, but $f^{-1}: y \to y^{1/3}$ is not differentiable at the origin.

(b) If U is non-empty and if $f\colon U \to V$ is a diffeomorphism, then E and F are isomorphic. In particular, their dimensions are equal, if they are finite. For the chain rule applied to $f^{-1} \circ f = \mathrm{id}_E$ implies that $Df^{-1}(f(u)) \circ Df(u) = \mathrm{id}_E$, if $u \in U$. Thus $Df(u) \in \mathscr{L}(E; F)$ has an inverse $Df^{-1}(f(u)) \in \mathscr{L}(F; E)$.

Definition 3.2

We say that $f\colon U \to V$ is slack and of class C^1 if f is of class C^1 and $Df(u) \in \mathscr{L}(E; F)$ is an isomorphism of E on F for all $u \in U$.

A diffeomorphism is thus well-behaved. The converse is false: $f\colon (\rho, \theta) \in \mathbf{R}^+ \times \mathbf{R} \to (\rho \cos \theta, \rho \sin \theta) \in \mathbf{R}^2 \setminus \{0\}$ is well-behaved, for:

$$\det Df(\rho, \theta) = \det \begin{pmatrix} \cos \theta & -\rho \sin \theta \\ \sin \theta & \rho \cos \theta \end{pmatrix} = \rho \neq 0$$

However, f is not a diffeomorphism, for it is not injective. The reason is that the open set $\mathbf{R}^+ \times \mathbf{R}$ is 'too big'; if we replace it by the open set $U = \mathbf{R}^+ \times \{0 < \theta < 2\pi\}$, the map f becomes a diffeomorphism of U on $f(U)$.

This leads to the following fundamental result.

3.2 THE INVERSE FUNCTION THEOREM

Theorem 3.1

Let U be a non-empty open subset of a Banach space E, and let f be a C^1 map of U to a Banach space F. If, for some $a \in U$, $Df(a)$ is an isomorphism of E on F, then there is an open neighbourhood I of a and an open neighbourhood J of $f(a)$ such that $f|_I : I \to J$ is a diffeomorphism. Moreover, the derivative of the inverse map f^{-1} at $y = f(x) \in J$ is given by $Df^{-1}(y) = [Df(x)]^{-1}$.

Remark

Put $b = f(a)$. Theorem 3.1 implies that the equation $f(x) = y$ has a unique solution x provided that y is chosen near enough to b and x is sought near enough to a.

3.2.1 The finite-dimensional case

We shall first prove the inverse function theorem when the dimensions of E and F are finite.

Step 1: With the notation of Theorem 3.1, note that we may assume $a=0$, $f(a)=0$, $E=F$ and $Df(a)=\mathrm{id}_E$: it is enough to replace $f(x)$ by $h(x)=[Df(a)]^{-1}[f(a+x)-f(a)]$.

Step 2: Since Df is continuous, given $\delta \in]0,1[$ there exists $\varepsilon > 0$ such that $\| u \| < \varepsilon$ implies that $\| Df(u) - Df(0) \| < \delta$. Take x and $x+y$ in the open ball $B(0,\varepsilon)$ with centre 0 and radius ε, and suppose that $f(x)=f(x+y)$. Since $Df(0)=\mathrm{id}_E$, the fundamental theorem of integral calculus (Theorem 2.10) implies that:

$$\begin{aligned} 0 = \| f(x+y) - f(x) \| &= \left\| \int_0^1 Df(x+ty)y \, dt \right\| \\ &= \left\| y + \int_0^1 [Df(x+ty) - Df(0)] \, y \, dt \right\| \\ &\geqslant \| y \| - \delta \| y \| = (1-\delta) \| y \|. \end{aligned} \tag{3.1}$$

Thus $y=0$ and f is injective on $B(0,\varepsilon)$.

Let us now show that the image of the ball $B(0,\varepsilon/2)$ under f contains a ball $B(0,r)$, provided that $0<r<\varepsilon(1-\delta)/4$.

Step 3: Let $z \in B(0,r)$ be given. The closed ball $\bar{B}(0,\varepsilon/2)$ of the *finite-dimensional* space E is compact. The continuous function $x \in \bar{B}(0,\varepsilon/2) \rightsquigarrow \| f(x) - z \|$ then attains its infimum at a point x_0.

We claim that x_0 is an interior point of this ball. If $\| x \| = \varepsilon/2$, the fundamental theorem of integral calculus and $Df(0)=\mathrm{id}_E$ imply that:

$$\begin{aligned} \| f(x) \| &= \left\| x + \int_0^1 [Df(tx) - Df(0)] \, x \, dt \right\| \geqslant \| x \| - \delta \| x \| \\ &= (1-\delta) \| x \| = \varepsilon(1-\delta)/2 > 2r. \end{aligned}$$

Thus $\| f(x) - z \| \geqslant \| f(x) \| - \| z \| > r > \| z \| = \| f(0) - z \| \geqslant \| f(x_0) - z \|$. The infimum therefore cannot be attained on the boundary of $\bar{B}(0,\varepsilon/2)$.

Step 4: Let us show that $f(x_0)=z$. Set $y = k[f(x_0)-z]$, where $k<0$ and $|k|$ is small enough for $\| x_0 + y \| < \varepsilon/2$ [this is possible, for $x_0 \in B(0,\varepsilon/2)$]. The fundamental theorem of integral calculus shows that

$$\| f(x_0+y) - f(x_0) - y \| = \left\| \int_0^1 [Df(x_0+ty) - Df(0)] \, y \, dt \right\| \leqslant \delta \| y \|$$

and hence:

$$\begin{aligned} \| f(x_0+y) - z \| &\leqslant \| f(x_0+y) - f(x_0) - y \| + \| f(x_0) - z + y \| \\ &\leqslant \delta \| y \| + (1+k) \| f(x_0) - z \| \\ &= [1 + k - \delta k] \, \| f(x_0) - z \|. \end{aligned}$$

Since $1+k-\delta k < 1$, the definition of x_0 shows that $f(x_0)=z$.

Step 5: Set $I = B(0,\varepsilon/2) \cap f^{-1}[B(0,r)]$. As f is continuous, this is an open neighbourhood of 0. By what is above, $f: I \to J = f(I)$ is a bijection. We shall prove that the inverse map is continuous.

If $y, y+k \in J$, there exist $x, x+k \in I$ such that $y = f(x)$ and $y+k = f(x+h)$. By equation (3.1) we have $\| f(x+h) - f(x) \| \geqslant (1-\delta) \| h \|$, and thus:

$$\| f^{-1}(y+k) - f^{-1}(y) \| \leqslant \| k \| / (1-\delta). \qquad \square$$

To sum up, $f: I \to J$ is a C^1 homeomorphism. It remains to prove that f^{-1} is differentiable and that its derivative at y is $\{Df[f^{-1}(y)]\}^{-1}$. We shall not do this here, for the proof is no more difficult in infinite dimensions and will be given in §3.2.2.

Remark

Step 4 above may be simplified. First we prove a lemma which will later be useful for other purposes.

Lemma 3.1

Let A be a subset of a normed v.s. E. If $f: A \to R$ is differentiable (see Definition 1.2) and has a minimum at an interior point a of A, then $Df(a) = 0$.

Proof. Let $h \in E$. Since there is an open set containing a and contained in A, $a + th \in A$ for $t \in \mathbf{R}$, if $|t|$ is small enough. The differentiable function $t \mapsto f(a+th) \in \mathbf{R}$ thus has a minimum at $t = 0$. Hence $0 = (\mathrm{d}/\mathrm{d}t) f(a+th)|_{t=0} = Df(a)h$, and $Df(a) = 0$, since h is arbitrary.

Since E is finite-dimensional, all norms on E are equivalent (see §A.2.5). We may thus assume that there is an inner product on E and that $\langle x, x \rangle$ defines the square of the norm. The function $x \mapsto \langle f(x) - z, f(x) - z \rangle$ therefore has a minimum when $x = x_0$. Its derivative at x_0 is zero. We thus obtain (see §1.2.8): $Df(x_0)[f(x_0) - z] = 0$. Since $Df(x_0)$ is invertible, $f(x_0) = z$.

Remark

The proof above has two drawbacks. First, it uses the local compactness of E, which is finite-dimensional (see J. Dieudonné, p. 106, for the converse: a locally compact normed space is finite-dimensional). Secondly, it does not give an algorithm which enables us to approximate the solution efficiently. The proof which follows is free of these defects.

3.2.2 Proof of the inverse function theorem

Now E and F are arbitrary Banach spaces. The proof requires several steps.

Lemma 3.2

(Inversion of an isomorphism between Banach spaces). The set $\mathrm{GL}(E; F)$ of all isomorphisms from E on to F is an open subset of $\mathscr{L}(E; F)$, and the map $\mathbf{J}: u \mapsto u^{-1}$ of $\mathrm{GL}(E; F)$ to $\mathrm{GL}(F; E)$ is continuous.

Proof. We may assume that $E = F$. For if $v \in \mathrm{GL}(E; F)$ (assumed non-empty for otherwise everything that follows would be trivial), the map $u \mapsto v^{-1} \circ u$ of $\mathscr{L}(E; F)$ in $\mathscr{L}(E; E)$ is continuous and $\mathrm{GL}(E; F)$ is simply the inverse image of $\mathrm{GL}(E; E)$ under this map.

Let $u \in \mathrm{GL}(E; E)$ and $h \in \mathscr{L}(E; E)$. We shall prove that $u + h \in \mathrm{GL}(E; E)$ if $\| h \| < 1/\| u^{-1} \|$. For simplicity, denote by 1 the identity map id_E. Since u is invertible and $u + h = u \circ [1 + u^{-1} \circ h]$, it is enough to prove (set $v = -u^{-1} \circ h$) that $1 - v$ is invertible if $\| v \| = \| u^{-1} \circ h \| \leqslant \| u^{-1} \| \, \| h \| < 1$. To do this the formula $(1 - x)^{-1} = 1 + x + x^2 + \ldots$, where $x \in \mathbf{R}$, $|x| < 1$, suggests consideration of the sequence $X_0 = 1$, $X_1 = 1 + v$, $\ldots$, $X_n = 1 + v + \ldots + v^n$. This is a Cauchy sequence in $\mathscr{L}(E; E)$, for (see §A.2.2) we have $\| X_{p+q} - X_p \| = \| v^{p+q} + \ldots + v^{p+1} \| \leqslant \| v \|^{p+1} + \ldots + \| v \|^{p+q}$, which tends to zero as $p \to +\infty$, since $\| v \| < 1$. Since $\mathscr{L}(E; E)$ is complete (see Theorem A.2), X_n converges to X. But $(1 - v) \circ X_n = 1 - v^{n+1}$, composition is continuous (see §A.3.1) and $v^{n+1} \to 0$ as $n \to +\infty$; thus $(1 - v) \circ X = 1$ and X is the desired inverse of $1 - v$.

Let us retain the notation and show that $\mathbf{J}$ is continuous:

$$\mathbf{J}(u + h) - \mathbf{J}(u) = [(1 - v)^{-1} - 1] \circ u^{-1} = [\lim_{n \to \infty} X_n - 1] \circ u^{-1}$$
$$= \lim_{n \to \infty} (v + \ldots + v^n) \circ u^{-1}.$$

Since

$$\| v + \ldots + v^n \| \leqslant \| v \| + \ldots + \| v \|^n \leqslant \| v \|/(1 - \| v \|)$$
$$\leqslant \| h \| \, \| u^{-1} \|/(1 - \| u^{-1} \circ h \|)$$

we have $\lim_{h \to 0} \| \mathbf{J}(u + h) - \mathbf{J}(u) \| = 0$. □

Proof of Theorem 3.1

Step 1: As in §3.2.1, we may assume that $a = 0$, $f(a) = 0$, $E = F$ and $Df(0) = \mathrm{id}_E$.

Step 2: Set $g(x) = x - f(x)$. We have $g(0) = 0$, $Dg(0) = 0$. Since Dg is continuous, there exists $r > 0$ such that if $x \in B(0, 2r)$ then $\| Dg(x) \| \leqslant 1/2$. The mean-value theorem (Theorem 2.3), applied in the ball $B(0, 2r)$, which is convex, shows that:

$$\| g(x) \| = \| g(x) - g(0) \| \leqslant \| x \|/2 < r$$

Hence $g[B(0, 2r)] \subset B(0, r)$.

Let $y \in B(0, r)$; we shall see that there exists a unique $x \in B(0, 2r)$ such that $f(x) = y$, that is, $h(x) = x$, if we put $h(x) = y + g(x)$.

Let us study h. If $x \in B(0, 2r)$, then $\| h(x) \| \leqslant \| y \| + \| g(x) \| \leqslant 2r$; h is thus a map of $B(0, 2r)$ into itself. On the other hand Theorem 2.3 applied to g in the ball $B(0, 2r)$, gives $\| h(u) - h(v) \| = \| g(u) - g(v) \| \leqslant \| u - v \|/2$. Thus h is a contraction of $B(0, 2r)$ into itself. By Banach's fixed-point theorem

(Appendix B), there exists a unique $x \in B(0, 2r)$ such that $h(x) = x$, that is $f(x) = y$. It follows that $f^{-1}: B(0, r) \to B(0, 2r)$ exists.

Step 3: We shall prove that f^{-1} is 2-Lipschitz, that is, that for all $x, y \in B(0, r)$, $\| f^{-1}(x) - f^{-1}(y) \| \leqslant 2 \| x - y \|$. By Step 2, we may write $x = f(u)$, $y = f(v)$, where u, $v \in B(0, 2r)$. With the definition $w = g(w) + f(w)$ of g, we obtain:

$$\| u - v \| \leqslant \| g(u) - g(v) \| + \| f(u) - f(v) \|$$

and, since $\| g(u) - g(v) \| \leqslant \| u - v \|/2$, $\| u - v \| \leqslant 2 \| f(u) - f(v) \|$; that is:

$$\| f^{-1}(x) - f^{-1}(y) \| \leqslant 2 \| x - y \|$$

Step 4: We shall prove that f^{-1} is of class C^1. Since $x \rightsquigarrow [Df(x)]^{-1}$ is the composition of the continuous map Df and $\mathbf{J}$ (see Lemma 3.2), it is continuous. We may therefore choose the number r from the beginning so small that $[Df(x)]^{-1}$ exists on $B(0, 2r)$; moreover, its continuity shows that there exists $K > 0$ such that $\| [Df(x)]^{-1} \| \leqslant K$ on $B(0, 2r)$. Having done this, let $y, y + k \in B(0, r)$; then $x = f^{-1}(y)$ and $x + h \in f^{-1}(y + h)$ are in $B(0, 2r)$. Thus

$$\begin{aligned} \| f^{-1}(y + k) - f^{-1}(y) - [Df(x)]^{-1}k \| &= \| (x + h) - x \\ &\quad - [Df(x)]^{-1}[f(x + h) - f(x)] \| \\ &= \| [Df(x)]^{-1}[f(x + h) - f(x) - Df(x)h] \| \\ &\leqslant K \| f(x + h) - f(x) - Df(x)h \|. \end{aligned}$$

But $\| f(x + h) - f(x) - Df(x)h \| / \| h \| \to 0$ as $\| h \| \to 0$ and, since f^{-1} is 2-Lipschitz, we have:

$$\| h \| / \| k \| = \| (x + h) - x \| / \| k \| = \| f^{-1}(y + k) - f^{-1}(y) \| / \| k \| \leqslant 2$$

It follows that f^{-1} is differentiable and that $Df^{-1}(y) = [Df(x)]^{-1}$. This may be written as $Df^{-1} = \mathbf{J} \circ Df \circ f^{-1}$, which is continuous, being the composition of continuous maps. □

Remark

The continuity of Df is essential. For f: $\mathbf{R} \to \mathbf{R}$, defined by:

$$f(x) = \begin{cases} (x/2) + x^2 \sin(1/x), & \text{if } x \neq 0 \\ 0, & \text{if } x = 0 \end{cases}$$

is differentiable and $f'(0) \neq 0$. However, there is no open interval I which contains 0 and on which f is invertible. (Prove this.)

Corollary 3.1

(Invariance of domain theorem) Let U be an open subset of a Banach space E, and let f: $U \to F$ be a slack mapping of class C^1 to a Banach space F. Then $f(U)$ is open in F.

Proof. This follows immediately from Theorem 3.1: by hypothesis $Df(u)$ is an isomorphism of E on F, for all u in U. □

3.3 THE IMPLICIT FUNCTION THEOREM

Let us study the function $f:\mathbf{R}^2 \to \mathbf{R}$ defined by: $f(x,y) = x^2 + y^2 - 1$. If $|a| \neq 1$, $b > 0$ and $f(a,b) = 0$, there exists an interval A which contains a and an interval B which contains b, such that to each $x \in A$ there corresponds a unique $y \in B$ such that $f(x,y) = 0$. This defines a function $x \in A \rightsquigarrow y = g(x) \in B$ such that $f(x, g(x)) = 0$. In the present case $g(x) = (1 - x^2)^{1/2}$.

We could have associated to the number a another number c, here equal to b, such that $f(a,c) = 0$. We would then obtain another function h, here equal to $-g$, such that $f(x, h(x)) = 0$. We say that each of these functions g and h is defined implicitly by the equation $f(x,y) = 0$.

If we had chosen $|a| = 1$, it would have been impossible to find such a function defined on an open interval containing a.

More generally, consider m equations $f_i(x,y) = 0$, $i = 1, \ldots, m$, in m unknowns $y = (y_1, \ldots, y_m)$, depending on n parameters $x = (x_1, \ldots, x_n)$. Assume that $f_i(a,b) = 0$, $i = 1, \ldots, m$, for $a = (a_1, \ldots, a_n)$ and $b = (b_1, \ldots, b_m)$. Under what conditions can we make correspond to each x near to a, a unique y near to b, satisfying the m equations $f_i(x,y) = 0$? The following theorem gives a simple criterion which enables us to answer this question.

Theorem 3.2

(Implicit function theorem) Let E and F be Banach spaces, let U be an open subset of E, V an open subset of F, and let f be a C^1 map of $U \times V$ in a Banach space G. Assume that at $(a,b) \in U \times V$ the partial derivative $D_2 f(a,b) \in \mathscr{L}(F; G)$ is an isomorphism of F on G. Then there exist a neighbourhood A of a, a neighbourhood W of $f(a,b)$ and a unique C^1 map $g_1 : A \times W \to V$ such that, for all $(x,w) \in A \times W$, we have $f(x, g_1(x,w)) = w$.

Proof. The map $\phi : U \times V \to E \oplus G$, defined by $\phi(x,y) = (x, f(x,y))$, is of class C^1 and its derivative at (a,b) is given by:

$$D\phi(a,b)(h,k) = (h, D_1 f(a,b)h + D_2 f(a,b)k)$$

for $h \in E$, $k \in F$. Since $D_2 f(a,b)$ is an isomorphism, $D\phi(a,b)$ is too, and has inverse $(h', k') \rightsquigarrow (h', D_2 f(a,b)^{-1}[k' - D_1 f(a,b)h'])$, for $h' \in E$, $k' \in G$.

By the inverse function theorem, ϕ is a diffeomorphism of a neighbourhood of (a,b) on a neighbourhood of $\phi(a,b)$. We may assume that this neighbourhood of $\phi(a,b)$ is the product of an open neighbourhood $A \subset U$ of a and an open neighbourhood W of $f(a,b)$. The inverse diffeomorphism ϕ^{-1} is plainly of the form $(x,w) \rightsquigarrow (x, g_1(x,w))$. Then $g_1 : A \times W \to V$ is the desired map. □

The corollary which follows answers the question put in the introduction.

Corollary 3.2

Suppose the hypotheses of Theorem 3.2 hold, and assume in addition that $f(a, b) = 0$. Then there exist a neighbourhood A of a, a neighbourhood B of b and a unique C^1 map $g\colon A \to B$ such that $f(x, g(x)) = 0$ for all $x \in A$. Moreover:

$$Dg(x) = -[D_2 f(x, g(x))]^{-1} \circ D_1 f(x, g(x))$$

Proof. It is enough to take $w = 0$, $g(x) = g_1(x, 0)$ and $B = g_1(A \times \{0\})$. The chain rule applied to $f(x, g(x)) = 0$ gives $D_1 f(x, g(x)) + D_2 f(x, g(x)) \circ Dg(x) = 0$. But since $D_2 f$ is continuous, $D_2 f(a, b)$ is invertible and (x, y) is near to (a, b), Lemma 3.2 shows that $D_2 f(x, g(x))$ is invertible. The last part of the corollary now follows. □

Remark

If $K = \mathbf{R}$ or $\mathbf{C}$, $E = K^n$, $F = K^m$ and $f = (f_1, \ldots f_m)$, the condition '$D_2 f(a, b)$ is invertible' means that the determinant of the $m \times m$ matrix, with i, jth element $D_{n+j} f_i(a, b) = (\partial f_i / \partial y_j)(a, b)$, is non-zero.

Theorem 3.3

Let E and F be Banach spaces, let U be an open subset of E, and let $f\colon U \to F$ be a C^1 map. Let us make the following hypotheses:

(a) $Df(a)$ is surjective for some $a \in U$;
(b) there is a closed subspace E_2 of E such that E is the direct sum of $E_1 = \ker Df(a)$ and E_2.

Then $f(U)$ contains an open neighbourhood of $f(a)$.

Proof. First note that, since $Df(a) \in \mathscr{L}(E; F)$ is continuous, its kernel E_1 is a closed subspace of E. It is thus a Banach space.

Next observe that hypothesis (b) is redundant if E is finite-dimensional or is a Hilbert space (by the projection theorem, we can take E_2 to be the orthogonal complement of E_1).

By the definition of E_2, the map $f\colon U \subset E_1 \oplus E_2 \to F$ has a continuous partial derivative $D_2 f(a)\colon E_2 \to F$ which is an isomorphism. The hypotheses of Theorem 3.2 are thus satisfied, and $f(U)$ contains the open set W mentioned in that theorem. □

3.4 AN APPLICATION OF THE IMPLICIT FUNCTION THEOREM

We shall show, by means of an example, how Theorem 3.3 can be used to prove the existence of solutions of differential equations.

Let E be the Banach space $C^1([0, 1])$ of all C^1 functions u: $[0, 1] \to \mathbf{R}$, normed by $\| u \|_{C^1} = \sup_{[0,1]} |u(t)| + \sup_{[0,1]} |u'(t)|$, studied in §A.1.3; let F be the Banach space $C^0([0, 1])$ of all continuous functions g: $[0, 1] \to \mathbf{R}$, endowed with the uniform convergence norm.

Define a map f: $E \to F$ by $f(u)$: $t \rightsquigarrow u'(t) + tu^2(t)$.

We have seen (Chapter 1) that f is of class C^1 and that $Df(0) = \mathrm{d}/\mathrm{d}t$. Thus $Df(0)$ is surjective: by §2.6.2 $g \in C^0([0, 1])$ is the image under $Df(0)$ of a primitive of g, from which we see that it is in $C^1([0, 1])$. Also $E_1 = \ker Df(0)$ is the set of constant functions. If E_2 is the closed subspace of E whose elements are the functions u with zero integral over $[0, 1]$, then clearly $E = E_1 \oplus E_2$.

By Theorem 3.3, there exists $\varepsilon > 0$ such that if $g \in C^0([0, 1])$ satisfies $|g(x)| < \varepsilon$ for all $x \in [0, 1]$, then there exists $u \in C^1([0, 1])$ such that $u'(t) + tu^2(t) = g(t)$ for all $t \in [0, 1]$.

Note

We shall return to the implicit function theorem and its corollaries in Chapter 8.

The interested reader will find in Appendix C an algorithm for construction of the root of an equation $f(x) = 0$ which considerably improves Banach's theorem used in the proof of the inverse function theorem.

Appendix D gives two *global* inverse function theorems.

4

Higher-order derivatives

This chapter is devoted to higher-order derivatives of a differentiable map, to the rules relating to them and to Taylor's formula.

4.1 SUCCESSIVE DERIVATIVES, AND SCHWARZ' THEOREM

Let E and F be normed v.s., let U be an open subset of E, and let f: $U \to F$ be differentiable. The derivative of f is a map of U to the normed v.s. $\mathscr{L}(E; F)$. We may then ask whether Df is differentiable at $a \in U$, or even at all points of U.

Definition 4.1

We say that f is *twice differentiable* at $a \in U$ if Df is differentiable at a. The derivative of Df at a is denoted by $D^2f(a)$. This is an element of $\mathscr{L}(E; \mathscr{L}(E: F))$, which is called the *second derivative of f at a.*

Of course the definition does not require that Df should exist at each point of U. It is enough that Df should be defined on an open neighbourhood $U' \subset U$ of a, and be differentiable at a.

If Df: $U \to \mathscr{L}(E; F)$ is differentiable at each point of U, we say that f is twice differentiable on U. If this is the case, $u \in U \rightsquigarrow D^2f(u)$ is a map D^2f: $U \to \mathscr{L}(E; \mathscr{L}(E; F))$ called the *second derivative* of f. It may happen that D^2f is continuous on U; if so, we say that f is *of class C^2 on U.*

The following properties are plainly equivalent:

(a) f is of class C^2 on U;
(b) f is differentiable on U and Df is of class C^1 on U.

4.1.1 Interpretation of $D^2f(a)$

Recall that $\mathscr{L}(E; \mathscr{L}(E; F))$ is canonically isomorphic to the normed v.s. $\mathscr{L}^2(E; F)$(see Theorem A.9). The image of $D^2f(a) \in \mathscr{L}(E; \mathscr{L}(E; F))$ under this isomorphism is thus a continuous bilinear map of $E \times E$ to F, which is defined by $(h,k)\mapsto(D^2f(a)h)k$: $D^2f(a)$ makes correspond to $h\in E$ the element $D^2f(a)h$ of $\mathscr{L}(E;F)$, and the image of $k\in E$ under this last map is $(D^2f(a)h)k$, which we shall write as $D^2f(a)(h,k)$.

4.1.2 Calculation of $D^2f(a)$

Suppose f is twice differentiable at a. Apply equation (1.1) to Df:

$$D^2f(a)h = D(Df)(a)h = \frac{\mathrm{d}}{\mathrm{d}t}Df(a+th)\,|_{t=0}.$$

We thus obtain an element of $\mathscr{L}(E;F)$, the value of which at $k\in E$ is:

$$D^2f(a)(h,k) = \frac{\mathrm{d}}{\mathrm{d}t}Df(a+th)k\,|_{t=0}.$$

But by equation (1.1):

$$Df(a+th)k = \frac{\mathrm{d}}{\mathrm{d}s}f(a+th+sk)\,|_{s=0}.$$

Thus

$$D^2f(a)(h,k) = \frac{\mathrm{d}}{\mathrm{d}t}\frac{\mathrm{d}}{\mathrm{d}s}f(a+th+sk)\,|_{t=s=0}.$$

Example: continuous bilinear maps

Consider again the continuous bilinear map b: $E_1\oplus E_2\to F$ of §1.2.3. If $a=(a_1,a_2)$ and $h=(h_1,h_2)$ are elements of $E_1\oplus E_2$, we have seen that $Db(a)h = b(h_1,a_2)+b(a_1,h_2)$. The map Db:

$$E_1\oplus E_2\to\mathscr{L}(E_1,E_2;F)$$

is thus a continuous linear map. By §1.2.2 its derivative exists and is constant. This constant is the element $D^2f(a)\in\mathscr{L}^2(E_1\oplus E_2;F)$, the value of which at the elements $h=(h_1,h_2)$ and $k=(k_1,k_2)$ of $E_1\oplus E_2$ is $D^2f(a)(h,k)=b(h_1,k_2)+b(k_1,h_2)$, as we also see by use of the above.

4.1.3 Schwarz' theorem

Theorem 4.1

If f: $U\subset E\to F$ is twice differentiable at $a\in U$, then $\|f(a+u+v)-f(a+u)-f(a+v)+f(a)-D^2f(a)(u,v)\|/(\|u\|+\|v\|)^2$ tends to zero as u and v tend to zero. In particular, $D^2f(a)$ is a bilinear symmetric form: $D^2f(a)(u,v)=D^2f(a)(v,u)$ for all $u,v\in E$.

Proof. Since Df is differentiable at a, given any $\varepsilon>0$, there exists $\delta>0$ such that if $\| x \| < 2\delta$, then:

$$\|Df(a+x) - Df(a) - D^2f(a)x\| \leqslant \varepsilon \| x \|. \tag{4.1}$$

Choose u and v with norm less than δ and set:

$$g_v(u) = f(a+u+v) - f(a+u) - f(a+v) + f(a) - D^2f(a)(u, v).$$

Using the rule for differentiation of the continuous linear map $u \leadsto D^2f(a)(u, v)$ we have:

$$\begin{aligned} Dg_v(u) &= Df(a+u+v) - Df(a+u) - D^2f(a)(u, v) \\ &= [Df(a+u+v) - Df(a) - D^2f(a)(u+v)] \\ &\quad - [Df(a+u) - Df(a) - D^2f(a)u] \end{aligned}$$

and so, by equation (4.1), $\|Dg_v(u)\| \leqslant 2\varepsilon(\| u \| + \| v \|)$.

Now apply Corollary 2.2 to g_v in the ball $B(0, \| u \|)$. Since $g_v(0) = 0$, we obtain:

$$\begin{aligned} \| g_v(u) \| = \| g_v(u) - g_v(0) \| &\leqslant 2\varepsilon(\| u \| + \| v \|)\| u \| \\ &\leqslant 2\varepsilon(\| u \| + \| v \|)^2 \end{aligned}$$

and the first part of the theorem follows.

Since $f(a+u+v) - f(a+u) - f(a+v) + f(a)$ is symmetric in u and v, so is $D^2f(u)(u, v)$. □

We shall now define derivatives of arbitrary order, retaining the same notation: f is a map of an open subset U of a normed v.s. E to a normed v.s. F.

Definitions 4.2

By the expression 'f is once differentiable' we shall mean that f is differentiable. We shall define, by recurrence on the integer $n > 0$, the expression 'f is n times differentiable on U' and the derivative $D^nf(a)$ of order n of f at $a \in U$.

To this end we recall that the normed v.s. of continuous linear maps of E to the normed v.s. $\mathscr{L}^{n-1}(E; F)$ of continuous $(n-1)$-linear maps of $E \times \ldots \times E$ $(n-1$ times) to F is canonically isomorphic to $\mathscr{L}^n(E; F)$(see Theorem A.9).

We say that f is n times differentiable at $a \in U$ if there is an open neighbourhood $U' \subset U$ of a in which f is $(n-1)$ times differentiable, and the map $u \in U' \leadsto D^{n-1}f(a)$ of U' to $\mathscr{L}^{n-1}(E; F)$ is differentiable at a. The derivative of $D^{n-1}f$ at a is denoted by $D^nf(a)$ and is called the derivative of order n of f at a. It is an element of $\mathscr{L}(E; \mathscr{L}^{n-1}(E; F)) = \mathscr{L}^n(E; F)$ and its value at $(h_1, \ldots, h_n) \in E^n$ is denoted by $D^nf(a)(h_1, \ldots, h_n)$. This is the same as $(D(D^{n-1}f)(a)\, h_1)(h_2, \ldots, h_n)$.

It is left to the reader to verify that this definition is equivalent to the following: f is n times differentiable at a if it is differentiable in an open neighbourhood $U' \subset U$ of a and Df: $U' \to \mathscr{L}(E; F)$ is $(n-1)$ times differentiable at a.

If f is n times differentiable at each point of U, we say that it is n times differentiable on U. In this case, $u \leadsto D^n f(u)$ defines a map $D^n f\colon U \to \mathscr{L}^n(E; F)$ which is called the derivative of order n of f. If the map $D^n f$ is continuous on U, we say that f is of class C^n on U. It is left to the reader to verify the equivalence of this definition with the following: f is of class C^n on U if f is of class C^1 on U and Df is of class C^{n-1} on U.

Note that since a map which is differentiable at a point is continuous at that point, an n times differentiable map is of class C^{n-1}.

4.1.4 The space $C^n(U; F)$

The linearity of differentiation shows that the C^n maps $f\colon U \to F$ form a v.s.; we denote this by $C^n(U; F)$. In particular, $C^0(U; F)$ is the space of continuous maps from U to F.

The intersection $\bigcap_{n>0} C^n(U; F)$ will be written as $C^\infty(U; F)$; its elements are maps $f\colon U \to F$ which are of class C^n for every integer $n > 0$.

We say that such an element is of class C^∞ or, what amounts to the same, is *infinitely differentiable,* that is, n times differentiable for all n.

4.1.5 Generalization of Schwarz' theorem

Theorem 4.2

If $f\colon U \subset E \to F$ is n times differentiable at $a \in U$, then $D^n f(a) \in \mathscr{L}^n(E; F)$ is a symmetric n-linear map. In other words, for all $h_1, \ldots, h_n \in E$ and every permutation s of the integers $\{1, 2, \ldots, n\}$, we have:

$$D^n f(a)(h_1, \ldots, h_n) = D^n f(a)(h_{s(1)}, \ldots, h_{s(n)})$$

Proof. The question arises only when $n \geqslant 2$, and Theorem 4.1 answers it when $n = 2$.

Thus we suppose that $n \geqslant 3$ and prove the theorem by induction, supposing it true for $n - 1$. By hypothesis, $D^n f(a) = D(D^{n-1}f)(a)$. On the other hand, $D^{n-1}f$ has values in the subspace $\mathscr{L}_s^{n-1}(E; F)$ of $\mathscr{L}^{n-1}(E; F)$ consisting of symmetric $(n-1)$-linear maps. Thus $D^n f(a)h_1 = D(D^{n-1}f)(a)h_1 \in \mathscr{L}_s^{n-1}(E; F)$ for $h_1 \in E$, and:

$$D^n f(a)(h_1, \ldots, h_n) = (D^n f(a)h_1)(h_2, \ldots, h_n)$$

is a symmetric function of $h_2, \ldots, h_n$.

If the permutation s leaves the number 1 fixed, the theorem is proved.

If the permutation s does not leave the number 1 fixed, it is the product of permutations leaving 1 fixed and a permutation which interchanges 1 and 2, leaving $3, \ldots, n$ fixed. The theorem will thus be proved if we can prove that $D^n f(a).(h_1, h_2, \ldots, h_n)$ does not change when h_1 and h_2 are interchanged. But

$D^n f(a) = D^2(D^{n-2}f(a))$; thus, by applying Theorem 4.1 to $D^{n-2}f$, we find:

$$(D^n f(a)\, h_1)h_2 = D^2(D^{n-2}f)(a)(h_1, h_2) = D^2(D^{n-2}f)(a)(h_2, h_1)$$
$$= (D^n f(a)h_2)h_1. \qquad \square$$

4.1.6 Calculation of $D^n f(a)$

Let $f: U \subset E \to F$ be n times differentiable at $a \in U$. Repeated application of equation (1.1) gives the following generalization of the result in §4.1.2:

$$D^n f(a)(h_1, \ldots, h_n) = \frac{\mathrm{d}}{\mathrm{d}t_n} \cdots \frac{\mathrm{d}}{\mathrm{d}t_1} f\left(a + \sum_{i=1}^{n} t_i h_i\right)\Bigg|_{t_1 = \ldots = t_n = 0}, \tag{4.2}$$

where $h_1, \ldots, h_n \in E$.

This formula reduces the calculation of $D^n f(a)$ to that of the successive derivatives of a function of n real variables and with values in F.

The number on the right-hand side of equation (4.2) generalizes the notion of the derivative of f in the direction of a vector h of E. Denote by $\mathrm{d}^n f(a)$ the operator which maps $(h_1, \ldots, h_n) \in E^n$ to this number on the right-hand side (we sometimes call this the Gâteaux derivative of order n of f at a). We shall now establish a generalization of Theorem 2.5.

Theorem 4.3

Assume that:

(a) $\mathrm{d}^n f(x)$ exists in an open neighbourhood U of a;
(b) $\mathrm{d}^n f(x) \in \mathscr{L}^n(E; F)$ for all x in U;
(c) $x \in U \rightsquigarrow \mathrm{d}^n f(x)$ is continuous.

Then f is of class C^n at a, and $\mathrm{d}^n f(a) = D^n f(a)$.

Proof. We proceed by induction on n. If $n = 1$, $t \rightsquigarrow f(x + th)$ is differentiable at $t = 0$ for all $x \in U$ and all $h \in E$, and its derivative $\mathrm{d}f(x)h$ depends continuously on x. Set $x = a + sh$, where $0 \leqslant s \leqslant 1$ and $\| h \|$ is so small that $x \in U$. Then $t \rightsquigarrow f(a + sh + th)$ is differentiable at $t = 0$ and its derivative $\mathrm{d}f(a + sh)h$, which is simply the derivative with respect to s of $s \rightsquigarrow f(a + sh)$, is continuous. The fundamental theorem of integral calculus thus shows that:

$$f(a + h) - f(a) = \int_0^1 \mathrm{d}f(a + sh)h \,\mathrm{d}s$$

Hence:

$$\| f(a + h) - f(a) - \mathrm{d}f(a)h \| = \left\| \int_0^1 [\mathrm{d}f(a + sh) - \mathrm{d}f(a)]\, h \,\mathrm{d}s \right\|$$

$$\leqslant \| h \| \int_0^1 \| \mathrm{d}f(a + sh) - \mathrm{d}f(a) \| \,\mathrm{d}s$$

which is $o(\|h\|)$ as $x \mapsto df(x)$ is continuous at a. Thus $Df(a)$ exists, it is equal to $df(a)$ and Df is therefore continuous at a.

Suppose the theorem has been proved for $n-1$. By hypothesis and by the fundamental theorem of integral calculus:

$$\begin{aligned}[D^{n-1}f(a+h_n) - D^{n-1}f(a)]\,(h_1,\ldots,h_{n-1}) &= [d^{n-1}f(a+h_n) - d^{n-1}f(a)]\,(h_1,\ldots,h_{n-1}) \\ &= \int_0^1 (d^n f(a+sh_n)\,h_n)(h_1,\ldots,h_{n-1})\,ds\end{aligned}$$

It follows that:

$$\begin{aligned}\|[D^{n-1}f(a+h_n) - D^{n-1}f(a) - d^n f(a)h_n]\,(h_1,\ldots,h_{n-1})\| &\leqslant \left\| \int_0^1 [d^n f(a+sh_n) - d^n f(a)]\,(h_1,\ldots,h_n)\,ds \right\| \\ &\leqslant \|h_1\| \ldots \|h_n\| \int_0^1 \|d^n f(a+sh_n) - d^n f(a)\|\,ds.\end{aligned}$$

Since $d^n f$ is continuous at a, the integral tends to zero with $\|h\|$. Thus $D^{n-1}f(a+h_n) - D^{n-1}f(a) - d^n f(a)h_n = o(\|h_n\|)$; hence $D^n f(a)$ exists and equals $d^n f(a)$, and is therefore continuous at a. □

4.2 RULES FOR CALCULATION

4.2.1 Continuous linear maps

If f is a continuous linear map of a normed v.s. E to a normed v.s. F, we know (§1.2.1) that its derivative is constant. Hence $D^n f = 0$ for $n \geqslant 2$ and f is of class C^∞.

4.2.2 Continuous bilinear maps

Let E_1, E_2 and F be normed v.s. and let $b\colon E_1 \times E_2 \to F$ be a continuous bilinear map. We know that Db is differentiable and that $D^2 b$ is a constant. Hence $D^n b = 0$ for $n \geqslant 3$ and b is of class C^∞.

4.2.3 Leibniz' rule

Let E, F_1, F_2 and G be normed v.s., let U be an open subset of E, let $f\colon U \to F_1$ and $g\colon U \to F_2$ be maps which are of class C^k and let $b\colon F_1 \times F_2 \to G$ be a continuous bilinear map. As in §1.2.8, define a map $p\colon U \to G$ by $x \mapsto b(f(x), g(x))$. Then p is of class C^k.

Proof. Suppose that f and g are of class C^1. Then p is differentiable at $a \in U$, and if $h \in E$ we have:

$$Dp(a)h = b(Df(a)h, g(a)) + b(f(a), Dg(a)h). \tag{4.3}$$

Thus Dp is continuous and p is of class C^1.

We shall prove the theorem by induction, supposing it true for $k-1$. Suppose that f and g are of class C^k. The map $v: \mathscr{L}(E; F_1) \times F_2 \to \mathscr{L}(E; G)$, which to each $A \in \mathscr{L}(E; F_1)$ and each $l \in F_2$ makes correspond the continuous linear map $h \rightsquigarrow b(A(h), l)$ of E to G, is continuous and bilinear. Moreover, Df and g are of class C^{k-1}. Since the theorem holds for $k-1$, it follows that $x \rightsquigarrow v(Df(x), g(x)) = b(Df(x), g(x))$ is of class C^{k-1}.

We can show in the same way that $x \rightsquigarrow b(f(x), Dg(x))$ is of class C^{k-1}. Hence, by equation (4.3), Dp is of class C^{k-1} and p is of class C^k. □

The proof shows that in the statement of the theorem the phrase 'of class C^k' may be replaced by 'k times differentiable'.

4.2.4 Higher-order derivatives of a product

If f and g are scalar-valued functions defined on an open set where they are of class C^n, the last rule shows that their product is of class C^n. Applying n times the rule for differentiation of a product, we see that the nth derivative is of the form:

$$(fg)^{(n)} = \sum_{q=0}^{n} A_{qn} f^{(n-q)} g^{(q)}$$

where the coefficients A_{qn} are constants independent of f and g. Choose $f(x) = e^{ax}$, $g(x) = e^{bx}$, where a and b are scalars. After simplification the preceding formula becomes $(a+b)^n = \sum A_{qn} a^{n-q} b^q$. Since a and b are arbitrary, the coefficients A_{qn} are those of the binominal formula: $A_{qn} = {}^nC_q = n!/(n-q)!q!$. Hence:

$$(fg)^{(n)} = \sum_{q=0}^{n} {}^nC_q f^{(n-q)} g^{(q)}. \tag{4.4}$$

The proof of the following generalization is left to the reader. Under the hypothesis of §4.2.3 the derivative of order n of p is given by:

$$D^n p(x)(h_1, \ldots, h_n) = \sum_{q=0}^{n} \sum b[D^q f(x)(h_{i_1}, \ldots, h_{i_q}), D^{n-q} g(x)(h_{j_1}, \ldots, h_{j_{n-q}})]$$

where $\sum$ is extended over the ${}^nC_q$ partitions of $(h_1, \ldots, h_n)$ of E^n into two subsets such that $i_1 < \ldots < i_q$ and $j_1 < \ldots < j_{n-q}$.

We can write this expression in a form similar to equation (4.4). Let sym be the operation of symmetrization, which to $A \in \mathscr{L}^n(E; F)$ makes correspond the symmetric n-linear map sym (A) defined by:

$$\text{sym}\,(A).(h_1, \ldots h_n) = \frac{1}{n!} \sum_s A[h_{s(1)}, \ldots, h_{s(n)}]$$

where the summation is extended over all permutations s of $\{1, 2, \ldots, n\}$. Taking account of Schwarz' theorem (Theorem 4.2), the preceding expression reduces to:

$$D^n p = \text{sym} \sum_{q=0}^{n} {}^nC_q b(D^{n-q}f, D^q g).$$

4.2.5 Derivatives of composite maps

Theorem 4.4

Let E, F and G be normed v.s., let U be an open subset of E, and let V be an open subset of F. Suppose that f: $U \to V$ and g: $V \to G$ are C^n maps. Then $g \circ f$ is also of class C_n.

Proof. We already know (§1.26) that if f and g are differentiable, then so is $g \circ f$, and $D(g \circ f)(x) = Dg[f(x) \circ Df(x)$.

If f and g *are of class* C^1, Dg and f are continuous; the composite map $x \rightsquigarrow Dg[f(x)]$ is therefore continuous; $g \circ f$ is thus of class C^1.

Let us prove the theorem by induction, and suppose that it is true for $n-1$. Assuming that f and g are of class C^n, Dg and f are of class C^{n-1} and, by the inductive hypothesis, the composite map $x \rightsquigarrow Dg[f(x)]$ is of class C^{n-1}. So is Df. Also the map $\mathscr{L}(F; G) \times \mathscr{L}(E; F) \to \mathscr{L}(E; G)$, which to $A \in \mathscr{L}(F; G)$ and $B \in \mathscr{L}(E; F)$ makes correspond their composition $A \circ B$, is continuous and bilinear. Thus by Leibniz' rule:

$$x \rightsquigarrow Dg[f(x)] \circ Df(x)$$

is of class C^{n-1} and $g \circ f$ is of class C^n. □

Remark

Under the hypothesis of Theorem 4.4, and with the notion of the tangent functor T introduced in §1.2.6, we have $T^n(g \circ f) = T^n g \circ T^n f$ by repeated application of the formula $T(g \circ f) = Tg \circ Tf$.

An analogous result in terms of D is decidedly more complicated:

$$D^n(g \circ f)(x)(h_1, \ldots, h_n) = \sum_{q=1}^{n} \sum D^q g[f(x)] \, (D^{r_1}f(x)(h_1^{(1)}, \ldots h_{r_1}^{(1)}), \ldots,$$
$$D^{r_q}f(x)(h_1^{(q)}, \ldots, h_{r_q}^{(q)}))$$

where the sum $\sum$ is extended over the $n!/(r_1! \ldots r_q!)$ partitions of $h_1, \ldots, h_n$ into q subsets, containing $r_1, \ldots, r_q$ elements, and in which the h are arranged in increasing order of their indices. If we write, for shortness (!),

$$(D^{r_1}f, \ldots, D^{r_q}f)(x)(h_1, \ldots h_n)$$
$$= (D^{r_1}f(x)(h_1, \ldots, h_{r_1}), D^{r_2}f(x)(h_{r_1+1}, \ldots, h_{r_1+r_2}), \ldots,$$
$$D^{r_q}f(x)(h_{r_1+\ldots+r_{q-1}+1}, \ldots, h_n)),$$

use of Schwarz' theorem and the symmetrization 'sym' enables the formula above to be written in the shorter form:

$$D^n(g\circ f) = \mathrm{sym} \sum_{q=1}^{n} \sum_{r_1+\ldots+r_q=n} \frac{n!}{r_1!\ldots r_q!}(D^q g\circ f)(D^{r_1}f,\ldots,D^{r_q}f).$$

(See L. E. Fraenkel, *Math. Proc. Camb. Phil. Soc.* **83** (1978), p. 159. See also H. Federer, *Geometric Measure Theory,* Springer-Verlag (1969), p. 222.)

4.2.6 Inversion of an isomorphism between Banach spaces

We shall use the notation of Lemma 3.2: GL(E; F) will stand for the open set formed by the automorphisms of a Banach space E on a Banach space F; $\mathbf{J}$ will denote the inversion $u \rightsquigarrow u^{-1}$ which maps GL(E; F) on GL(F; E).

Theorem 4.5

The map $\mathbf{J}$ is of class C^∞ and $D\mathbf{J}(u)h = -u^{-1}\circ h\circ u^{-1}$ for $h\in\mathscr{L}(E; F)$.

Proof. Set $Lh = -u^{-1}\circ h\circ u^{-1}$. Clearly L is linear and it is continuous for, by §A.2.2, $\| Lh \| \leqslant \| u^{-1} \| \; \| u \| \; \| h \|$. We study $A = \mathbf{J}(u+h) - \mathbf{J}(u) - Lh$. We have $A = (u+h)^{-1}\circ[1-(u+h)\circ u^{-1} + (u+h)\circ u^{-1}\circ h\circ u^{-1}] = (u+h)^{-1}\circ h\circ u^{-1}\circ h\circ u^{-1}$, so that $\| A \| \leqslant \| (u+h)^{-1} \| \; \| u^{-1} \|^2 \; \| h \|^2$. Since $\mathbf{J}$ is continuous (Lemma 3.2), $(u+h)^{-1}\to u^{-1}$ if $h\to 0$, and thus $\| A \|/\| h \| \to 0$. This shows that $\mathbf{J}$ is differentiable and that $D\mathbf{J}(u)h = -u^{-1}\circ h\circ u^{-1}$.

Now we show that $\mathbf{J}$ is of class C^1. To do this we introduce a piece of notation: if $a, b\in\mathscr{L}(F; E)$ we denote by $f(a,b)$ the linear map $h\rightsquigarrow -a\circ h\circ b$ of $\mathscr{L}(E; F)$ to $\mathscr{L}(F; E)$. Thus $D\mathbf{J}(u) = f(u^{-1}, u^{-1})$. We see that:

$$(a,b)\in\mathscr{L}(F; E)\times\mathscr{L}(F; E) \rightsquigarrow f(a,b)\in\mathscr{L}[\mathscr{L}(E; F); \mathscr{L}(F; E)]$$

is bilinear and continuous, for $\| f(a,b)h \| = \| a\circ h\circ b \| \leqslant \| a \| \; \| b \| \; \| h \|$. Hence $D\mathbf{J}$ is continuous, for it is the composition of the continuous map $u \rightsquigarrow (u^{-1}, u^{-1})$ of GL(E; F) to $\mathscr{L}(F; E)\times\mathscr{L}(F; E)$ and of the continuous map $(a,b)\rightsquigarrow f(a,b)$. This shows that $\mathbf{J}$ is of class C^1.

Suppose that the theorem has been proved up to order n, and assume that $\mathbf{J}$ is of class C^n, $n\geqslant 1$. The map $u\rightsquigarrow u^{-1}$ is thus of class C^n. Also the bilinear continuous map $(a,b)\rightsquigarrow f(a,b)$ is of class C^∞, by §4.2.2. It follows from Leibniz' rule that $D\mathbf{J}$: $u\rightsquigarrow f(u^{-1}, u^{-1})$ is of class C^n; hence the map $\mathbf{J}$ is of class C^{n+1}. This shows that $\mathbf{J}$ is of class C^∞. □

Remark

Let $a\in$ GL(E; F) and $h\in\mathscr{L}(E; F)$ be such that $\| a^{-1}\circ h \| < 1$. The formula:

$$(a+h)^{-1} = \sum_{k=0}^{\infty} (-a^{-1}\circ h)^k\circ a^{-1}$$

proved in Lemma 3.2 enables us to show that (exercise):

$$D^k\mathbf{J}(a)(h_1,\ldots,k_k)=\sum_s(-1)^k(a^{-1}\circ h_{s(1)}\circ a^{-1})\circ\ldots\circ(a^{-1}\circ h_{s(k)}\circ a^{-1})$$

where the summation is over all permutations s of $\{1, 2, \ldots, k\}$.

4.2.7 Inverse of a C^n diffeomorphism

Theorem 4.6

Let $f\colon U\to V$ be a C^n diffeomorphism, $n\geqslant 1$, from an open subset U of a normed v.s. E to an open subset V of a normed v.s. F. Then the inverse diffeomorphism f^{-1} is of class C^n.

Proof. If $n=1$ the result reduces to the definition; there is nothing to prove and we know that $Df^{-1}=[Df\circ f^{-1}]^{-1}$. This can also be written as $Df^{-1}=\mathbf{J}\circ Df\circ f^{-1}$.

Suppose the theorem holds for $n-1\geqslant 1$. To prove it for n, we shall show that if f is of class C^n, then Df^{-1} is of class C^{n-1}. Since f is, in particular, of class C^{n-1} the inductive hypothesis shows that f^{-1} is of class C^{n-1}. Since f is of class C^n, Df is of class C^{n-1}. As for $\mathbf{J}$, we have just seen that it is of class C^∞. By Theorem 4.4 and the formula $Df^{-1}=\mathbf{J}\circ Df\circ f^{-1}$, we conclude that Df^{-1} is of class C^{n-1}. □

Corollary 4.1
If in the inverse function theorem (Theorem 31) we assume in addition that f is of a class C^n, $n\geqslant 1$, then f is a local diffeomorphism of class C^n. (Its local inverse f^{-1} is of class C^n.)

4.2.8 Map defined on a direct sum

Suppose that E is the direct sum $E_1\oplus\ldots\oplus E_n$ of normed v.s., and let f be a map of an open subset U of E to a normed v.s. F.

Suppose that f is of class C^k, $k\geqslant 1$. By using the results in §1.2.9 and Theorem 4.4 about composite maps, we see that each partial derivative $D_rf\colon U\to\mathscr{L}(E_r; F)$ exists and is of class C^{k-1}. Moreover, if $h=(h_1,\ldots,h_n)$, $h_r\in E_r$, we know that:

$$Df(a)h=\sum_{r=1}^{n}D_rf(a)h_r \text{ for } a\in U. \tag{4.5}$$

We intend to find a similar formula for $D^kf(a)$, first of all for $k=2$. To this end we use the formula:

$$D^2f(a)(h^{(1)},h^{(2)})=\frac{\mathrm{d}}{\mathrm{d}t}\frac{\mathrm{d}}{\mathrm{d}s}f(a+th^{(1)}+sh^{(2)})|_{t=s=0}$$

for $h^{(1)}, h^{(2)} \in E$. For small enough t, $a + th^{(1)} \in U$ and we have:

$$\frac{\mathrm{d}}{\mathrm{d}s} f(a + th^{(1)} + sh^{(2)})|_{s=0} = Df(a + th^{(1)})h^{(2)}.$$

By equation (4.5), applied at the point $a + th^{(1)}$, this may be written as:

$$\sum_{r=1}^{n} D_r f(a + th^{(1)})h_r^{(2)}.$$

Differentiating with respect to t and taking $t = 0$, we obtain:

$$D^2 f(a)(h^{(1)}, h^{(2)}) = \sum_{r=1}^{n} \frac{\mathrm{d}}{\mathrm{d}t} D_r f(a + th^{(1)})|_{t=0}\, h_r^{(2)}. \tag{4.6}$$

Let us evaluate the right-hand side. The map $x \in E \leadsto D_r f(x) \in \mathscr{L}(E_r; F)$ is differentiable and its derivative $D(D_r f)(x) \in \mathscr{L}(E; \mathscr{L}(E_r; F))$ may be determined, in its turn, by means of the formula:

$$D(D_r f)(x)h^{(1)} = \sum_{s=1}^{n} D_s(D_r f)(x)h_s^{(1)}$$

where $D_s(D_r f)(x) \in \mathscr{L}(E_s; \mathscr{L}(E_r; F))$. Combining this result with $(\mathrm{d}/\mathrm{d}t)D_r f(a + th^{(1)})|_{t=0} = D(D_r f)(a)h^{(1)}$ and equation (4.6) we obtain:

$$D^2 f(a)(h^{(1)}, h^{(2)}) = \sum_{r,s=1}^{n} D_s(D_r f)(a)(h_s^{(2)}, h_r^{(1)}). \tag{4.7}$$

We put $D_s(D_r f)(a) = D^2_{sr} f(a)$; this is an element of $\mathscr{L}(E_s; \mathscr{L}(E_r; F))$, which we know from Theorem A.9 to be canonically isomorphic to $\mathscr{L}(E_s, E_r; F)$. We call it a partial derivative of order 2 of f at a.

By Schwarz' theorem, $D^2 f(a)$ is a symmetric bilinear form. Equation (4.7) thus shows that:

$$\sum_{r,s} D^2_{sr} f(a)(h_s^{(1)}, h_r^{(2)}) = \sum_{r,s} D^2_{rs} f(a)(h_s^{(2)}, h_r^{(1)}).$$

Interchanging the summation indices r and s on the right-hand side, we obtain:

$$\sum_{r,s} D^2_{sr} f(a)(h_s^{(1)}, h_r^{(2)}) = \sum_{r,s} D^2_{rs} f(a)(h_r^{(2)}, h_s^{(1)}).$$

Since the $h_s^{(1)}$ and $h_r^{(2)}$ are arbitrary, it follows that

$$D^2_{rs} f(a)(h_s^{(1)}, h_r^{(2)}) = D^2{}_{rs} f(a)(h_r^{(2)}, h_s^{(1)}).$$

The map $D^2_{rs} f(a) \in \mathscr{L}(E_r, E_s; F)$ is thus the composition of the map:

$$(h_r^{(2)}, h_s^{(1)}) \in E_r \times E_s \leadsto (h_s^{(1)}, h_r^{(2)}) \in E_s \times E_r$$

and the map $D^2_{sr} f(a) \in \mathscr{L}(E_s, E_r; F)$. In particular, $D^2_{rr} f(a)$ is a symmetric bilinear map. However, one should not think that $D^2_{rs} f(a) = D^2_{sr} f(a)$: these are not even elements of the same space if $s \neq r$.

If f is of class $C^k, k \geqslant 2$, equation (4.7) is easily generalized:

$$D^k f(a)(h^{(1)}, \ldots, h^{(k)}) = \sum_{r_1, \ldots, r_k} D^k_{r_1, \ldots, r_k} f(a)(h^{(1)}_{r_1}, \ldots, h^{(k)}_{r_k}) \quad (4.8)$$

with an obvious notation.

4.2.9 Case where $E = \mathbf{R}^n$ or $\mathbf{C}^n$

Let us take $E_1 = \ldots = E_n = \mathbf{R}$(or $\mathbf{C}$) in the last paragraph. Then $\mathscr{L}(E_r; F)$ may be canonically identified with F, by Theorem A.8; thus $\mathscr{L}(E_s; \mathscr{L}(E_r; F))$ may be identified with $\mathscr{L}(E_s; F)$, that is, with F again. It follows from the last paragraph that $D_{sr}f(a)$ and $D_{rs}f(a)$ may be identified with the same element of F. *In this particular case* we may thus write $D_{sr}f(a) = D_{rs}f(a)$, often written

$$\frac{\partial^2 f}{\partial x^r \, \partial x^s}(a).$$

If, in addition, $F = \mathbf{R}^m$ (or $\mathbf{C}^m$), and if f is given by its components in the canonical basis $[f(x) = (f_1(x), \ldots, f_m(x))]$, the components of the bilinear form $D^2 f(a)$, defined on $\mathbf{R}^n \times \mathbf{R}^n$ and with values in $\mathbf{R}^m$, are $(\partial^2 f/\partial x^r \, \partial x^s)(a)$, $i = 1, \ldots, m$, $1 \leqslant r, s \leqslant n$, in canonical bases. More generally, the components of $D^k f(a)$ in the canonical bases are $D^k_{r_1, \ldots, r_k} f_i(a)$, often written $(\partial^k f_i / \partial x^{r_1} \ldots \partial x^{r_k})(a)$, $i = 1, \ldots, m$, $1 \leqslant r_1, \ldots, r_k \leqslant n$.

4.3 TAYLOR'S FORMULA

We intend to extend to 'order n' the fundamental theorem of integral calculus:

$$f(a+h) - f(a) = \int_0^1 Df(a+th)h \, \mathrm{d}t$$

which uses the first derivative of f.

Lemma 4.1

If u is an $(n+1)$-times differentiable function of a real variable t with values in a Banach space F, then:

$$D[u(t) + (1-t)Du(t) + \ldots + (1/n!)(1-t)^n D^n u(t)] = (1/n!)(1-t)^n D^{n+1} u(t).$$

Proof. We apply Leibniz's rule (§1.2.8) with $E = R$, $F_1 = R$, $F_2 = F$, $f(t) = (1-t)^k$, $g(t) = D^k u(t)$, $0 \leqslant k \leqslant n$ and, as a continuous bilinear map $b: F_1 \times F_2 = R \times F \to F$ the product of $r \in R$ by $y \in F$. The proof follows by 'telescoping'.

Lemma 4.2

If u is a C^{n+1} function of a real variable t, defined on an open set containing

$[0,1]$ and with values in a Banach space F, then:

$$u(1) - u(0) - u'(0) - \tfrac{1}{2}u''(0) - \ldots - (1/n!)u^{(n)}(0) = \int_0^1 ((1-t)^n/n!)u^{(n+1)}(t)\,dt$$

where $u^{(k)}(t) = D^k u(t)1 \in F$ is the derivative vector of order k.

Proof. We apply the fundamental theorem of integral calculus to the function:

$$t \rightsquigarrow [u(t) + (1-t)\,Du(t) + \ldots + (1/n!)(1-t)^n\, D^n u(t)]\,1$$

which is of class C^1, and use Lemma 4.1. □

Let us now weaken the hypotheses of Lemma 4.2.

Lemma 4.3

Let u be an $(n+1)$-times differentiable function of a real variable t, defined on an open set containing $[0,1]$, with values in a Banach space F, and such that there is a constant C majorizing $\|D^{n+1}u(t)\|$ for $0 \leqslant t \leqslant 1$. Then:

$$\|u(1) - u(0) - u'(0) - \ldots - (1/n!)u^{(n)}(0)\| \leqslant C/(n+1)!$$

Proof. We apply Theorem 2.1 with $[a,b] = [0,1]$,

$$f(t) = u(t) + (1-t)u'(t) + \ldots + (1/n!)(1-t)^n u^{(n)}(t)$$

and $g(t) = -C(1-t)^{n+1}/(n+1)!$

Lemma 4.1 enables us to satisfy the hypotheses of that theorem:

$$\|f'(t)\| = \|(1/n!)(1-t)^n u^{(n+1)}(t)\| \leqslant (C/n!)(1-t)^n = g'(t).$$

It follows that $\|f(1) - f(0)\| \leqslant g(1) - g(0)$, which is the inequality required. □

4.3.1 Taylor's formula with integral remainder

Theorem 4.7

Let f be a C^{n+1} map of an open subset U of a normed v.s. E in a Banach space F. If the segment $[a, a+h]$ is contained in U, then:

$$f(a+h) - f(a) - Df(a)h - \ldots - (1/n!)D^n f(a)(h)^n = \int_0^1 (1/n!)(1-t)^n D^{n+1} f(a+th)(h)^{n+1}\,dt$$

where $D^k f(a)(h)^k = D^k f(a)(\underbrace{h,\ldots,h}_{n \text{ times}})$.

Proof. Suppose that $[a, a+h] = \{a + th \colon 0 \leqslant t \leqslant 1\}$ lies in U. By Theorem

4.4 about composite maps, the function $u(t) = f(a + th)$ is of class C^{n+1} for $0 \leqslant t \leqslant 1$, and its derivative of order k is, as we see by induction on k:

$$u^{(k)}(t) = D^k u(t).1 = D^k f(a + th)(h)^k.$$

It is now enough to use Lemma 4.2. □

4.3.2 Taylor's formula with Lagrange's remainder

Theorem 4.8

In the last theorem let us replace the hypothesis 'f is of class C^{n+1}' by the following hypothesis: f is $(n + 1)$ times differentiable and there is a constant C which majorizes $\| D^{n+1}f(x) \|$ for all $x \in U$. Then:

$$\| f(a + h) - f(a) - Df(a)h - \ldots - (1/n!)\, D^n f(a)(h)^n \| \leqslant (C/(n + 1)!) \| h \|^{n+1}$$

Proof. Introduce the function $u(t) = f(a + th)$. It is $(n + 1)$ times differentiable and:

$$\| u^{(n+1)}(t) \| = \| D^{n+1}f(a + th)(h)^{n+1} \| \leqslant \| D^{n+1}f(a + th) \| \, \| h \|^{n+1} \leqslant C \| h \|^{n+1}.$$

Now use Lemma 4.3.

Theorem 4.9

Let $f\colon U \subset E \to F$ be an $(n - 1)$ times differentiable map of an open subset U of a normed v.s. E to a Banach space F. If f is n times differentiable at $a \in U$, then:

$$\| f(a + h) - f(a) - Df(a)h - \ldots - (1/n!)\, D^n f(a)(h)^n \| = o(\| h \|^n).$$

Proof (see H. Cartan). For $n = 1$, this is nothing more than the definition of the derivative at a. We shall prove the theorem by induction on n, assuming it true for $n - 1$. Consider the map:

$$g(h) = f(a + h) - f(a) - \ldots - (1/n!)\, D^n f(a)(h)^n$$

and determine its derivative. To do this look for the derivative of $\phi\colon h \rightsquigarrow D_k f(a)(h)^k$, where $k = 1, \ldots, n$. Since $D^k f(a) \in \mathscr{L}^k(E; F)$, if $l \in E$ we have:

$$D\phi(h)l = \frac{\mathrm{d}}{\mathrm{d}t} D^k f(a)(h + tl)^k \big|_{t=0} = D^k f(a)(l, h, \ldots, h)$$
$$+ D^k f(a)(h, l, h, \ldots, h) + \ldots + D^k f(a)(h, \ldots, h, l)$$

Since $D^k f(a)$ is symmetric, by Schwarz' theorem, all the terms in the last sum are equal. Thus $D\phi(h)l = kD^k f(a)(l, h, \ldots, h)$. If we write $D^k f(a)(h)^{k-1}$ for the continuous linear map $l \rightsquigarrow D^k f(a)(l, h, \ldots, h)$, it follows that:

$$Dg(h) = Df(a + h) - Df(a) - \ldots - (1/(n - 1)!) D^n f(a)(h)^{n-1}.$$

Application of the inductive hypothesis to Df shows that $\| Dg(h) \| = o(\| h \|^{n-1})$; given any $\varepsilon > 0$, there is therefore a $\delta > 0$ such that $\| Dg(h) \| \leqslant \varepsilon \| h \|^{n-1}$ if $\| h \| < \delta$. By the mean-value theorem we obtain $\| g(h) \| = \| g(h) - g(0) \| \leqslant \varepsilon \| h \|^n$, and so $\| g(h) \| = o(\| h \|^n)$. □

The same result could have been deduced from the last theorem, but at the cost of too strong a hypothesis: 'f has a derivative of order $n+1$ which is bounded on a neighbourhood of a'.

The estimate of Theorem 4.9 gives an nth-order generalization of the inequality which defines the first-order derivative:

$$\| f(a+h) - f(a) - Df(a)h \| = o(\| h \|).$$

It says that:

$$I = f(a) + Df(a)h + \ldots + (1/n!)\, D^n f(a)(h)^n$$

approximates $f(a+h)$ in a neighbourhood of a to order n in $\| h \|$. We may ask if the property characterizes I. The next section gives a first answer.

4.3.3 Uniqueness of Taylor's formula

Theorem 4.10

Let E and F be Banach spaces, and let f be a C^n map of an open subset U of E to F. Suppose there are symmetric maps $A_k \in \mathscr{L}^k(E; F)$, $k = 0, \ldots, n$, such that for some $a \in U$:

$$f(a+h) - \sum_{k=0}^{n} (1/k!) A_k (h)^k = o(\| h \|^n).$$

Then $D^k f(a) = A_k$ for $k = 0, \ldots, n$.

Proof. Put $B_k = (1/k!)(A_k - D^k f(a))$. Taylor's formula (§4.3.2) and the assumed relation give, by subtraction term by term:

$$\sum_{k=0}^{n} B_k (h)^k = o(\| h \|^n). \tag{4.9}$$

Letting $h \to 0$ in equation (4.9) we obtain $B_0 = 0$. Suppose that $B_0 = \ldots = B_k = 0$ for $k < n$. Then equation (4.9) gives:

$$-B_{k+1}(h)^{k+1} = \sum_{i=k+2}^{n} B_i (h)^i + o(\| h \|^n).$$

Thus:

$$\| B_{k+1} \| \; \| h \| \|^{k+1} \leqslant \sum_{i=k+2}^{n} \| B_i \| \; \| h \|^i + o(\| h \|^n).$$

Dividing both sides by $\| h \|^{k+1}$ and letting $h \to 0$, we obtain $B_{k+1} = 0$. All the B_i are thus zero. □

This result often enables us to identity easily the successive derivatives of f *when we know that f is of class C^n.*

Not only is the Taylor series to order n unique, but its property of approximating $f(a+h)$ to $o(\| h \|^n)$ almost characterizes it. This is shown by the following theorem, due to R. Abraham and J. Robbin (*Transversal Mappings and Flows,* Benjamin, 1967).

4.3.4 Converse of Taylor's formula

Theorem 4.11

Let f be a map from an open subset U of a normed v.s. E to a normed v.s. F. We make the following hypotheses:

(a) There are continuous maps a_j: $U \to \mathscr{L}^j(E; F)$, $j = 0, 1, \ldots, n$, such that $a_j(x)$ is symmetric for all x in U.

(b) Let $x \in U$ and $h \in E$ be such that $x + h$ lies in an open ball with centre x and contained in U. Put:

$$R_n(x, h) = f(x+h) - \sum_{k=0}^{n} (1/k!) a_k(x)(h)^k \tag{4.10}$$

where $(h)^k = \underbrace{(h, \ldots, h)}_{k \text{ times}}$, and assume that for all $x_0 \in U$, $\| R_n(x,h) \| / \| h \|^n \to 0$ if $(x, h) \to (x_0, 0)$.

Then f is of class C^n and $D^k f = a_k$ for $k = 0, 1, \ldots, n$.

Proof. (E. Nelson) *The case* $n = 1$. By hypothesis, $f(x+h) = a_0(x) + a_1(x)h + R_1(x,h)$. Since $\| R_1(x, h) \| = o(\| h \|)$, we have $a_0(x) = f(x)$; thus $f(x+h) = f(x) + a_1(x)h + o(\| h \|)$. Hence $a_1 = Df$ and, since a_1 is continuous, f is of class C^1.

We shall next prove the theorem for an arbitrary n by assuming it true for $n-1$. Since $(1/n!)a_n(x)(h)^n + R_n(x,h) = R_{n-1}(x,h)$ is such that:

$$\| R_{n-1}(x,h) \| / \| h \|^{n-1} \leqslant [(1/n!) \| a_n(x) \| + \| R_n(x,h) \| / \| h \|^n] \| h \| \to 0$$

as $h \to 0$, the inductive hypothesis implies that $a_0 = f, \ldots, a_{n-1} = D^{n-1} f$. Now *fix* $x \in U$ and choose $\varepsilon > 0$ so small that the ball with centre x and radius 2ε lies in U. Take $y, z \in E$ so that $\| y \|, \| z \| < \varepsilon$ and write $f(x+y+z)$ in two different ways:

$$f(x+y+z) = f(x+y) + Df(x+y)z + \ldots + (1/(n-1)!)D^{n-1}f(x+y)(z)^{n-1} + (1/n!)a_n(x+y)(z)^n + R_n(x+y, z)$$

and

$$f(x+y+z) = f(x) + Df(x)(y+z) + \ldots + (1/(n-1)!)D^{n-1}f(x)(y+z)^{n-1} + (1/n!)a_n(x)(y+z)^n + R_n(x, y+z).$$

By subtraction it follows that

$$g_0(y) + g_1(y)z + \ldots + g_{n-1}(y)(z)^{n-1} + [(1/n!)a_n(x+y)(z)^n - (1/n!)a_n(x)(z)^n + R_n(x+y, z) - R_n(x, y+z)] = 0 \quad (4.11)$$

where, using the symmetry of the n-linear form $a_n(x)$:

$$g_{n-1}(y)(z)^{n-1} = (1/(n-1)!)[D^{n-1}f(x+y) - D^{n-1}f(x) - a_n(x)y](z)^{n-1}. \quad (4.12)$$

Now suppose that $\| z \| < \| y \|$. The quantity in square brackets in equation (4.11) satisfies

$$\| [\ ] \| / \| y \|^n \leqslant (1/n!) \| a_n(x+y) - a_n(x) \| + \| R_n(x+y, z) \| / \| z \|^n + 2^n \| R_n(x, y+z) \| / \| y+z \|^n.$$

Since a_n is continuous, the first term on the right-hand side tends to zero as $y \to 0$. Since $(x+y, z) \to (x, 0)$ if $y \to 0$, hypothesis (b) shows that the second term tends to zero as $y \to 0$. This is also so for the third term. Thus from equation (4.11) we see that, if $y \to 0$:

$$\| g_0(y) + g_1(y)z + \ldots + g_{n-1}(y)(z)^{n-1} \| / \| y \|^n \to 0.$$

Taking $z = 0$ we obtain $g_0(y) = o(\| y \|^n)$, and thus:

$$g_1(y)z + \ldots + g_{n-1}(y)(z)^{n-1} = o(\| y \|^n).$$

Replace z by $z/2$ in this, then multiply both sides by 2 and subtract the relation obtained from the original relation; finally multiply both sides by 2. We obtain $g_2(y)(z)^2 + \ldots + (2 - (1/2^{n-3}))g_{n-1}(y)(z)^{n-1} = o(\| y \|^n)$.

Repeat this procedure so as to eliminate $g_2(y)(z)^2$, etc. We finally obtain:

$$g_{n-1}(y)(z)^{n-1} = o(\| y \|^n).$$

By equation (4.12) this implies that $\| D^{n-1}f(x+y) - D^{n-1}f(x) - a_n(x)y \| / \| y \| \to 0$ as $y \to 0$. By the definition of a derivative, this shows that $D^n f(x) = a_n(x)$ and, since a_n is continuous, f is of class C^n. □

4.3.5 An application

The above result is a convenient criterion for deciding whether a map is of class C^n. Consider, for example, the inversion $\mathbf{J}$: $u \rightsquigarrow u^{-1}$ introduced in §4.2.6. If $x \in \mathrm{GL}(E; E)$, $h \in \mathscr{L}(E; E)$ and $\| x^{-1} \circ h \| < 1$, we may write

$$\mathbf{J}(x+h) = \sum_{k=0}^{n} (-x^{-1} \circ h)^k \circ x^{-1} + R_n(x, h), \text{ where } R_n(x, h) = \sum_{k=n+1}^{\infty} (-x^{-1} \circ h)^k \circ x^{-1}.$$

Set:

$$A_k(x)(h_1, \ldots, h_k) = \sum_s (-1)^k (x^{-1} \circ h_{s(1)} \circ x^{-1}) \circ \ldots \circ (x^{-1} \circ h_{s(k)} \circ x^{-1})$$

where $h_1, \ldots, h_k \in \mathscr{L}(E; F)$ and the summation is over all permutations s of $\{1, 2, \ldots, k\}$.

Clearly $A_k(x) \in \mathscr{L}^k(E; F)$ is symmetric and $x \leadsto A_k(x)$ is continuous, since $x \leadsto x^{-1}$ is continuous. Also $R_n(x, h)$ is of the form $\rho_n(x, h)(h)^n$, where $\rho_n(x, h) \in \mathscr{L}^n(E; F)$ depends continuously on (x, h) and satisfies $\rho_n(x, 0) = 0$. Since $\mathbf{J}(x+h) = \sum_{k=0}^{n} A_k(x)(h)^k + R_n(x, h)$, Theorem 4.11 shows that $\mathbf{J}$ is of class C^n and that $D^k\mathbf{J} = A_k$, $k = 0, \ldots, n$, for all n.

4.4 TAYLOR SERIES AND ANALYTICITY

Definitions 4.3

Suppose f is of class C^∞. We may thus form the series $\sum_0^\infty (1/k!)D^kf(a)(h)^k$. We call this the Taylor series of f at a (if $a = $ o we sometimes call it the Maclaurin series of f).

If, given a number $r > 0$, the Taylor series converges for all $h \in E$ with $\| h \| < r$, and if its sum is equal to $f(a+h)$, we say that f is analytic at a.

Since $f(a+h) = \sum_0^n (1/k!)D^kf(a)(h)^k + R_n(a, h)$ for every integer $n > 0$, f is analytic at a if, and only if, the remainder $R_n(a, h)$ tends to zero with $1/n$.

Without dwelling on a study of analytic maps, which itself could occupy a whole book, we can nevertheless observe that the Taylor series may converge without having a sum equal to $f(a+h)$. Here are some examples which are interesting in their own right.

4.4.1 Non-analytic functions of class C^∞

Consider the function f: $\mathbf{R} \to \mathbf{R}$ defined by:

$$f(x) = \begin{cases} \mathrm{e.e}^{-1/x^2} & \text{if } x > 0 \\ 0 & \text{otherwise} \end{cases}$$

We shall show that it is C^∞. Clearly there is no problem for $x < 0$, as then $D^nf(x=0$ for all n. If $x > 0$, we see by induction on n that $D^nf(x) = x^{-3n}Q_n(x)f(x)$, where $Q_0(x) = 1$, $Q_1(x) = 2$, $Q_2(x) = 4 - 6x^2$ and $Q_n(x)$, for $n \geqslant 1$, is a polynomial of degree $2n - 2$ which satisfies the recurrence relation:

$$Q_{n+1}(x) = (2 - 3nx^2)Q_n(x) + x^3Q_n'(x).$$

The function f is thus C^∞ for $x > 0$.

We shall show, by induction on n, that $D^nf(0) = 0$. This is clear if $n = 0$. Suppose that this result has been established for n. Then:

$$\lim_{x \to 0^+} D^nf(x)/x = \lim_{x \to 0^+} x^{-3n-1}Q_n(x)\mathrm{e}^{-1/x^2} = 0$$

and thus $D^{n+1}f(0) = 0$. In particular, the nth derivative of f is everywhere continuous, even at $x = 0$. The function f is thus C^∞.

The Maclaurin series of f plainly converges to zero since $D_n f(0) = 0$. The sum is therefore not $f(h)$; f is not analytic at zero.

Numerous analogous functions may be obtained from f. Since f is C^∞ and increasing for $x > 0$, the function $g(x) = f[1 - f(x)]$ is of class C^∞, decreasing on R and such that $g(x) = 1$ for $x \leqslant 0$, $g(x) = 0$ for $x \geqslant 1$.

Let $a < b$ and put $h(x) = g(a - x)g(x - b)$. This function is C^∞, zero outside $]a - 1, b + 1[$, positive elsewhere, and equal to 1 on $[a, b]$. We call this a 'flat' function. (Exercise: draw its graph.)

5

The exponential function, and linear differential equations with constant coefficients

This chapter is devoted to the exponential of an endomorphism A of a Banach space.

We know that the usual exponential function $t \in \mathbf{R} \rightsquigarrow e^{at}$, where $a \in \boldsymbol{R}$, satisfies the differential equation $y' = ay$. This property extends to the map $t \in \mathbf{R} \rightsquigarrow \exp(tA)$. It is the key to the solution of linear differential equations with constant coefficients.

Throughout this chapter E will be a real or complex Banach space. For simplicity we shall write $\mathscr{L}(E; E) = \mathrm{End}(E)$, $\mathrm{GL}(E; E) = \mathrm{GL}(E)$ and $A \circ B = AB$ if $A, B \in \mathrm{End}(E)$. Lastly, if no ambiguity is possible, we shall put $\mathrm{id}_E = 1$.

5.1 DEFINITIONS OF THE EXPONENTIAL

Definition

We are motivated by the well-known property of the usual exponential: $e^x = \sum_0^\infty (x^n/n!)$ for $x \in \boldsymbol{R}$. If $A \in \mathrm{End}(E)$ we form the sequence $S_n = 1 + A + \ldots + (A^n/n!)$.

Theorem 5.1

The sequence S_n converges uniformly on each bounded subset of End(E) and its limit $\exp(A)$ satisfies $\| \exp(A) \| \leqslant \exp(\| A \|)$.

Proof. Let p and q be positive integers. By §§A.2.1 and A.2.2:

$$\| S_{p+q} - S_p \| \leqslant [\| A \|^{p+1}/(p+1)!] + \ldots + [\| A \|^{p+q}/(p+q)!] \quad (5.1)$$

which tends to zero as $p, q \to \infty$, for it is the truncated remainder of the convergent series $\sum_{0}^{\infty} \| A \|^n/n!$. The sequence S_n is thus a Cauchy sequence and it converges in the complete space End(E) to a limit, denoted by $\exp(A)$.

Let q tend to infinity in equation (5.1). If $\| A \| \leqslant R =$ constant, it follows that $\| \exp(A) - S_p \| \leqslant \sum_{p+1}^{\infty} R^k/k!$ and the convergence is uniform on the ball in End(E) with centre 0 and radius R.

Lastly $\| S_n \| \leqslant 1 + \ldots + \| A \|^n/n!$ and so $\| \exp(A) \| \leqslant \exp(\| A \|)$. □

Second Definition 5.2

For all $A \in \mathrm{End}(E)$ we have $\exp(A) = \lim_{n \to \infty} [1 + (A/n)]^n$.

Proof. By the binomial formula:

$$[1 + (A/n)]^n = \sum_{k=0}^{n} {}^n\mathrm{C}_k (1/n^k) A^k$$

where ${}^n\mathrm{C}_k = n!/[k!(n-k)!]$. Note that $(1/k!) - ({}^n\mathrm{C}_k/n^k) \geqslant 0$. It follows that:

$$\| \exp(A) - [1 + (A/n)^n] \| = \| \sum_{0}^{n} [(1/k!) - ({}^n\mathrm{C}_k/n_k)] A_k + \sum_{n+1}^{\infty} (A^k/k!) \|$$

$$\leqslant \sum_{0}^{n} [(1/k!) - ({}^n\mathrm{C}_k/n^k)] \| A \|^k + \sum_{n+1}^{\infty} (\| A \|^k/k!)$$

$$= \exp(\| A \|) - [1 + (\| A \|/n)]^n \to 0$$

as $n \to \infty$, by the classical formula $\mathrm{e}^x = \lim_{n \to \infty} [1 + (x/n)]^n$ for $x \in \mathbf{R}$.

5.2 PROPERTIES OF THE EXPONENTIAL

5.2.1 Continuity

Theorem 5.2

The map 'exp' is continuous; it is even locally Lipschitz: $\| \exp(A) - \exp(B) \| \leqslant \mathrm{e}^M \| A - B \|$ if M majorizes $\| A \|$ and $\| B \|$.

Proof. From §§A.2.1 and A.2.2 and from the identity:

$$A^n - B^n = A^{n-1}(A - B) + A^{n-2}(A - B)B + \ldots + A(A - B)B^{n-2} + (A - B)B^{n-1}$$

it follows that $\| A^n - B_n \| \leqslant n[\max(\| A \|, \| B \|)]^{n-1} \| A - B \|$.
The theorem follows. □

We shall prove later that 'exp' is of class C^∞.

5.2.2 Commutation with conjugation.

Theorem 5.3

If B is an isomorphism of E, and $A \in \mathrm{End}(E)$, we have $\exp(B^{-1}AB) = B^{-1}\exp(A)B$.

Proof. The conjugation $X \rightsquigarrow B^{-1}XB$ is continuous and commutes with $X \rightsquigarrow X^n$ for all integers $n \geqslant 0$.

5.2.3 Exponential of a sum of operators which commute

Theorem 5.4

Let $A, B \in \mathrm{End}(E)$ be such that $AB = BA$. Then $\exp(A+B) = \exp(A)\exp(B)$. In particular $\exp(A) \in \mathrm{GL}(E)$ and its inverse is $\exp(-A)$.

Proof. By Definition 5.2, and since $AB = BA$, we have:

$$\begin{aligned}\exp(A)\exp(B) &= \lim_{n\to\infty} [1 + (A/n)]^n \lim_{n\to\infty} [1 + (B/n)]^n \\ &= \lim_{n\to\infty} \{[1 + (A/n)][1 + (B/n)]\}^n.\end{aligned}$$

Set $v = [1 + (A/n)][1 + (B/n)]$, $u = 1 + (A+B)/n$ and use the estimate of §5.2.1:

$$\begin{aligned}\| v^n - u^n \| &\leqslant n[\max(\| u \|, \| v \|)]^{n-1} \| v - u \| \\ &\leqslant [1 + (\| A \| + \| B \|)/n + \| AB \|/n^2]^{n-1}(\| AB \|/n).\end{aligned}$$

If n is large enough, $\| AB \| \leqslant n(\| A \| + \| B \|)$, and then $[\ ]^{n-1} \leqslant [1 + (2\| A \| + 2\| B \|)/n]^n$, which tends to $\exp 2(\| A \| + \| B \|)$ as $n \to \infty$. Hence $\lim_{n\to\infty} \| v^n - u^n \| = 0$, and so

$$\exp(A)\exp(B) = \exp(A+B)$$

Noting that $\exp(0) = 1$, we see that $\exp(-A)$ is the inverse of $\exp(A)$.

The result is false if $AB \neq BA$ (take $E = \mathbf{R}^2$, A to be the matrix

$$\begin{pmatrix} 0 & 1 \\ 0 & 0 \end{pmatrix} \text{and } B \text{ the matrix} \begin{pmatrix} 0 & 0 \\ 1 & 0 \end{pmatrix}.$$

Remark

Every exponential is a square, for $\exp(A) = [\exp(A/2)]^2$.

Theorem 5.5

Let E be a finite-dimensional Banach space. Denote by $\det(A)$ and $\mathrm{tr}(A)$ the determinant and the trace of $A \in \mathrm{End}(E)$. Then $\det[\exp(A)] = \exp[\mathrm{tr}(A)]$.

Proof. Choose a basis of E. The determinant of A is a polynomial in the coefficients of its matrix and thus is a continuous map. Definition 5.2 shows that $\det[\exp(A)] = \det \lim_{n\to\infty} [1+(A/n)]^n = \lim_{n\to\infty} \{\det[1+(A/n)]\}^n$.

We study $\det[1+(A/n))$ when n is large. To do this, view the $d\times d$ matrix of $B \in \mathrm{End}(E)$ as a point in E^d, thinking of each line $B_1, \ldots, B_d$ as an element of E. Then det: $E^d \to \mathbf{R}$ (or $\mathbf{C}$) is a continuous d-linear map. By §1.2.3 it is differentiable, and if $h = (h_1, \ldots, h_d) \in E^d$ we have:

$$D(\det)(B)h = \sum_{k=1}^{d} \det(B_1, \ldots, B_{k-1}, h_k, B_{k+1}, \ldots, B_d).$$

In our case $B = 1$, $h = A/n$, and so:

$$\begin{aligned}\det[1+(A/n)] &= \det 1 + D(\det)(1)A/n + \mathrm{o}(1/n)\\ &= 1 + (1/n)\mathrm{tr}(A) + \mathrm{o}(1/n).\end{aligned}$$

Hence

$$\begin{aligned}\det[\exp(A)] &= \lim_{n\to\infty} [1+(1/n)\mathrm{tr}(A) + \mathrm{o}(1/n)]^n\\ &= \lim_{n\to\infty} [1+(1/n)\mathrm{tr}(A)]^n = \exp[\mathrm{tr}(A)].\end{aligned}$$

□

Corollary 5.1

If E is a real, finite-dimensional Banach space, then $\det[\exp(A)] > 0$. In particular, Exp: $\mathrm{End}(E) \to \mathrm{GL}(E)$ is not surjective. □

Theorem 5.6

Let A be an endomorphism of a Banach space E. The map f: $R \to \mathrm{GL}(E)$ defined by $f(t) = \exp(tA)$ is of class C^∞ and $f'(t) = Af(t)$ for all t.

Proof. Let us first show that f is differentiable at zero. Since $f(0) = \mathrm{id}_E$, we have:

$$\begin{aligned}\| f(s) - f(0) - sA \| &\leqslant \sum_{n=2}^{\infty} |s|^n \| A \|^n/n! =\\ &\exp(|s|\,\|A\|) - 1 - |s|\,\|A\| = \mathrm{o}(|s|).\end{aligned}$$

Hence $Df(0)$ exists and $f'(0) = Df(0).1 = A$.

Clearly tA and sA commute, for all real numbers s and t. By §5.2.3 we thus have $f(t+s) = f(t)f(s)$. The right-hand side is differentiable in s for $s = 0$, the left-hand side also has this property and $f'(t) = f(t)f'(0) = f(t)A$. Since A and $\exp(tA)$ commute, this may be written as $f'(t) = Af(t)$. Hence f is of class C^1.

The relation $f' = Af$ shows that if f is of class C^n so is f'; f is therefore of class C^∞. □

5.3 ONE-PARAMETER GROUP OF LINEAR AUTOMORPHISMS

Definition 5.3

By a one-parameter group of linear automorphisms of a Banach space E we shall mean a homomorphism h from the additive group of real numbers to the group GL(E). If the map h is continuous, we say that the group is continuous.

Example

f: $t \rightsquigarrow \exp(tA)$, where $A \in \mathrm{End}(E)$. For $f(0) = \mathrm{id}_E$ and, by Theorem 5.6, $f(s+t) = f(s)f(t)$ for all s, $t \in \mathbf{R}$.

Let us show that the above example is the general case.

Theorem 5.7

Every continuous one-parameter group h of linear automorphisms of a Banach space E is of the form $t \rightsquigarrow \exp(tA)$, where $A \in \mathrm{End}(E)$. There is thus one-to-one correspondence between these groups and endomorphisms A. We say that A is the generator of the group $\exp(tA)$.

Proof. Since h is continuous we may integrate both sides of $h(s+t) = h(s)h(t)$ between 0 and $a > 0$. We make the change of variable $s = u - t$ on the left-hand side and use property d of §2.6.4; we obtain:

$$\int_t^{t+a} h(u)\,\mathrm{d}u = h(t)\int_0^a h(s)\,\mathrm{d}s.$$

Since h is continuous we may choose $a > 0$ close enough to 0 for $L = a^{-1}\int_0^a h(s)\,\mathrm{d}s$ to be near to $h(0) = 1$. It follows from Lemma 3.2 that L is invertible; since L evidently commutes with $h(t)$, the above equality may be written as $h(t) = L^{-1} \circ \int_t^{t+a} h(u)\,\mathrm{d}u$. By §2.6.5 the right-hand side is differentiable with respect to t; thus so is h and, if we put $C = L^{-1}.\ [h(a) - 1]$, we obtain $h'(t) = Ch(t)$, where C commutes with $h(t)$. This enables us to find the derivative of $g(t) = h(t) \circ \exp(-Ct)$. With the aid of Theorem 5.6 we find:

$$g'(t) = h'(t)\exp(-Ct) + h(t)[-C\exp(-Ct)] = 0$$

By Theorem 2.4, g is constant and equal to $g(0) = h(0)\exp(0) = \mathrm{id}_E$. Thus $h(t) = \exp(Ct)$. □

Examples

(a) When $A = \mathrm{id}_E$ we obtain the group $t \rightsquigarrow e^t\mathrm{id}_E$ of homotheties with centre 0 and positive ratio (see Figure 5.1).

(b) When $A \in \mathrm{End}(\mathbf{R}^2)$ and has matrix representation $\begin{pmatrix} 0 & -1 \\ 1 & 0 \end{pmatrix}$, we

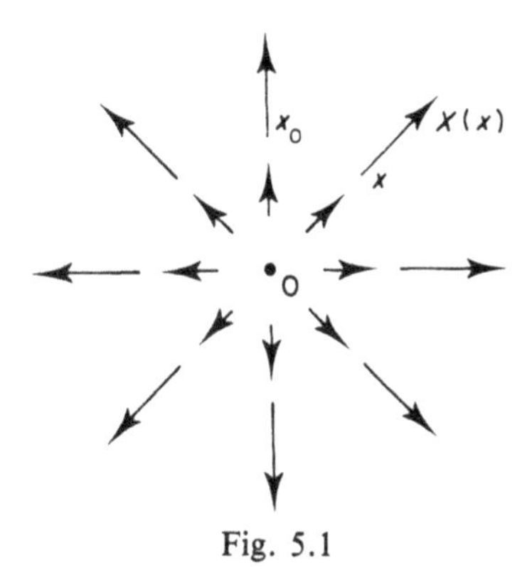

Fig. 5.1

have $A^2 = -\mathrm{id}_E$. It follows that $\exp(tA)$ has matrix representation $\begin{pmatrix} \cos t & -\sin t \\ \sin t & \cos t \end{pmatrix}$. We obtain the group of rotations about 0 as centre (Figure 5.2).

(c) When $A \in \mathrm{End}(\mathbf{R}^2)$ is represented by the matrix $\begin{pmatrix} 0 & 1 \\ 1 & 0 \end{pmatrix}$ we have $A^2 = \mathrm{id}_E$. It follows that $\exp(tA)$ is represented by the matrix $\begin{pmatrix} \mathrm{ch}\, t & \mathrm{sh}\, t \\ \mathrm{sh}\, t & \mathrm{ch}\, t \end{pmatrix}$. We obtain the group of hyperbolic rotations (Figure 5.3).

In each of these cases the diagrams show the orbit $\{\exp(tA)x_0 : t \in \mathbf{R}\}$ of a point x_0 of $E = \mathbf{R}^2$. If we interpret t as time, the velocity X at time t depends only on the position $x = \exp(tA)x_0$ at this instant:

$$X(x) = \frac{\mathrm{d}}{\mathrm{d}t} \exp(tA)x_0 = A \exp(tA)x_0 = Ax.$$

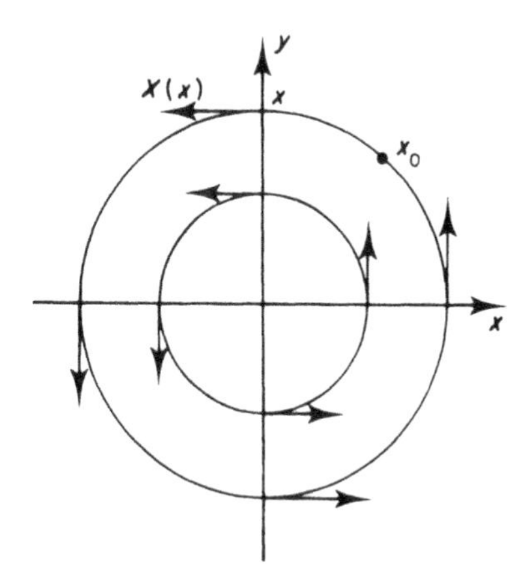

Fig. 5.2

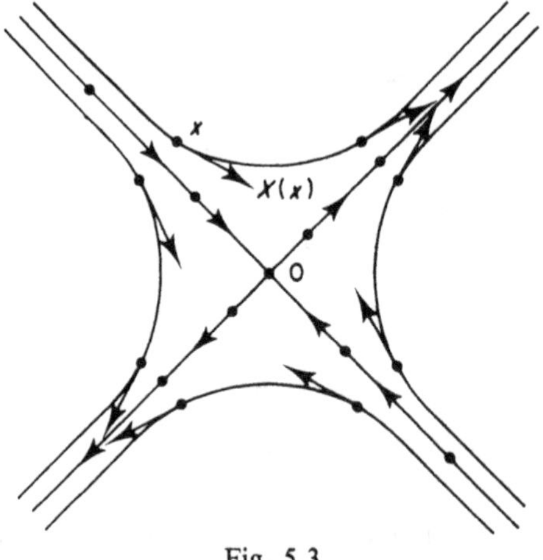

Fig. 5.3

5.4 HOMOGENEOUS LINEAR DIFFERENTIAL EQUATIONS WITH CONSTANT COEFFICIENTS

Definition 5.4

A homogeneous first-order differential equation with constant coefficients is an expression of the form:

$$\frac{dx}{dt} = Ax \tag{5.2}$$

where A is an endomorphism of a Banach space E. A solution of this equation is a differentiable map f from an interval I of $\mathbf{R}$ to E which satisfies $f'(t) = Af(t)$ for all $t \in I$.

There is thus a one-to-one correspondence between the set of equations we have just defined and the set of endomorphisms A of E.

Example

If $E = K^n$ ($K = \mathbf{R}$ or $\mathbf{C}$), the equation $x' = Ax$ is equivalent to a system of n scalar equations in n unknowns:

$$\frac{dx_i}{dt} = \sum_{j=1}^{n} a_i^j x_j, \; i = 1, 2, \ldots, n.$$

Remarks

(a) Denote by S_I the set of solutions $f: I \to E$ of equation (5.2). It is clear that if $f, g \in S_I$ and k is a scalar, then $f + g \in S_I$ and $kf \in S_I$. The set of solutions defined on the same interval I is therefore a vector space. If S_I is finite-dimensional, we shall have solved equation (5.2)—that is, found all its solutions—if we can exhibit a basis of S_I. Such a basis is called *a fundamental system of solutions.*

(b) A solution f is continuous, for it is assumed to be differentiable. Hence $f' = Af$ is continuous, and f is thus of class C^1. By induction it follows that f is of class C^∞.

5.4.1 Fundamental theorem

Theorem 5.8

Let E be a Banach space, $A \in \mathrm{End}(E)$, $x_0 \in E$, I an interval (open or closed, bounded or unbounded) in $\mathbf{R}$, and let $t_0 \in I$. Then there is a unique differentiable map $f\colon I \to E$ such that $\mathrm{d}f/\mathrm{d}t = Af(t)$ for all $t \in I$ and $f(t_0) = x_0$. In fact:

$$f(t) = \exp[(t - t_0)A]\, x_0.$$

We say that f is the solution on I of $x' = Ax$ which satisfies the initial condition $f(t_0) = x_0$.

Proof. Define a map $f\colon I \to E$ by $f(t) = \exp[(t - t_0)A]\, x_0$. Clearly $f(t_0) = x_0$, and Theorem 5.6 shows that f satisfies $x' = Ax$. Thus existence is established.

We shall now prove uniqueness. If g is a second candidate, $y = f - g$ satisfies $y'(t) = Ay(t)$ for all $t \in I$, and $y(t_0) = 0$. Integrate from t_0 to $s \in I$, using property (d) of §2.6.4:

$$y(s) = y(s) - y(t_0) = \int_{t_0}^{s} Ay(t)\,\mathrm{d}t = A\left[\int_{t_0}^{s} y(t)\,\mathrm{d}t\right].$$

We deduce that $\| y(s) \| \leqslant \| A \| \int_{t_0}^{s} \| y(t) \|\,\mathrm{d}t$, which enables us to estimate the derivative of $F(s) = \exp(-\| A \|\, s) \int_{t_0}^{s} \| y(t) \|\,\mathrm{d}t$. We find that $F'(s) \leqslant 0$ for all $s \in I$. Since $F(s) \geqslant 0$ and $F(t_0) = 0$, if follows that $F(s) = 0$, and thus $y(s) = 0$ for all $s \in I$. Hence $f = g$. □

Remarks

(a) Take $I = \mathbf{R}$. Then there exists a solution f of $x' = Ax$ defined for all $t \in \mathbf{R}$ and satisfying $f(t_0) = x_0$, t_0 and x_0 being arbitrarily fixed.

Now if I is an arbitrary interval in $\mathbf{R}$ which contains t_0, it is clear that the restriction $f|_I$ of f to I satisfies the differential equation $x' = Ax$ and the initial condition $f|_I(t_0) = x_0$. By uniqueness it coincides with the solution of §5.4.1. We express this by saying that the solutions $t \rightsquigarrow f(t)$ of an equation $x' = Ax$ may be extended to the whole of $\mathbf{R}$. We also say that the maximal (non-extendable) solutions of $x' = Ax$ are defined on the whole of $\mathbf{R}$.

(b) For each $t_0 \in I$ the map $f \rightsquigarrow f(t_0) = x_0$ from the vector space S_I of solutions of $x' = Ax$ to the space E is a vector space isomorphism. Indeed, the map which takes $x_0 \in E$ to $\exp[(t-t_0)A]\,x_0$ is injective (uniqueness of solutions), surjective (existence of solutions) and plainly linear. It follows that the dimension of S_I is equal to that of E.

(c) Once we have obtained what we call the general solution $\exp(tA)x$ of the equation $x' = Ax$, we might believe that the problem is over. This is not so, for the series $\exp(tA) = \sum(tA)^n/n!$ may be unsuitable for calculations. If we keep the solution in this form, there are certainly questions which are difficult to answer: is $\exp(tA)x$ periodic? Does it remain bounded as $t \to +\infty$? We shall show that it is possible to replace $\exp(tA)$ by a simpler expression, at least if E is finite-dimensional.

5.5 EXPLICIT CALCULATION OF SOLUTIONS

Throughout this section, E is a space of finite-dimension n.

5.5.1 The endomorphism A is diagonalizable

In this case E has a basis $\{e_k\}$ of eigenvectors of A, with corresponding eigenvalues $a_k : Ae_k = a_k e_k$. It follows that $A^n(e_k) = (a_k)^n e_k$ for all n, and hence $\exp(tA)e_k = \exp(ta_k)e_k$.

The functions $\exp(ta_k)e_k$ are thus solutions of $x' = Ax$. As the number of these is $n = \dim E$, and they are linearly independent, by Remark (b) above they form a basis of the space of solutions.

In practice, once the eigenvalues a_k have been determined, the e_k are sought by the method of indeterminate coefficients, using the fact that $\exp(ta_k)e_k$ satisfies $x' = Ax$.

5.5.2 The endomorphism A is nilpotent

Recall that $A \in \text{End}(E)$ is nilpotent with index N if $A^N = 0$ while $A^p = 0$ for $p = 1, 2, \ldots, N-1$. If this is so, $\exp(tA)$ reduces to the polynomial $1 + tA + \ldots + (tA)^{N-1}/(N-1)!$. The general solution $\exp(tA)x$ of $x' = Ax$ will then be a vector with components which are polynomials in t of degree less than N. We find the components of this vector by the method of indeterminate coefficients.

Example

The set E of polynomials in u with coefficients in **R** and of degree less than n is a vector space of dimension n on **R**. As differentiation with respect to u reduces the degree of a polynomial, $p' \in E$ if $p \in E$. The map D, which takes each polynomial $p \in E$ to its derivative p', is thus an endomorphism of E. As the nth derivative of a polynomial of degree less than n is zero, D is nilpotent with index n. Thus:

$$\exp(tD) = 1 + tD + \ldots + (tD)^{n-1}/(n-1)!$$

Let us identify $\exp(tD)$ by operating on $p \in E$ by the right-hand side; we obtain:

$$p(u) + tp'(u) + \ldots + [t^{n-1}/(n-1)!]\, p^{(n-1)}(u)$$

By Taylor's formula this is simply $p(u+t)$. hence $\exp(tD)$ is the translation operator $T_t : p \leadsto p(t + \circ)$.

5.5.3 General case on the field C

Let E be a complex n-dimensional vector space. Then by the d'Alembert–Gauss theorem, the characteristic polynomial $p(\lambda) = \det(A - \lambda.1)$ of $A \in \operatorname{End}(E)$ factorizes:

$$p(\lambda) = (\lambda - \lambda_1)^{r_1} \ldots (\lambda - \lambda_m)^{r_m}.$$

The endomorphism A has n eigenvalues $\lambda_1, \ldots \lambda_m$ counting these according to their multiplicities $r_1, \ldots, r_m$.

Appendix E has a proof of the following result. Let E_i be the kernel of $(A - \lambda_i.1)^{r_i}$. Thus E_i is of dimension r_i, and E is the direct sum of the E_i. Let us draw some conclusions from this result.

(a) $A(x) \in E_i$ if $x \in E_i$. For $(A - \lambda_i.1)^{r_i} A(x) = A[(A - \lambda_i.1)^{r_i} x] = 0$. The operator A thus leaves E_i invariant and induces an operator $A_i = A\,|_{E_i}$ on E_i.

(b) $A = \sum A_i$, the A_i commute, $A_i(E_k) = 0$ if $i \neq k$, and $(A_i - \lambda_i.1)^{r_i} = 0$. The proof is obvious.

We may now find explicitly the general solution $\exp(tA)x$ of the equation $x' = Ax$. Since $A = \sum A_i$ and the A_i commute, by §5.2.3 we have:

$$\exp(tA) = \exp(tA_1) \ldots \exp(tA_m).$$

Decompose $x \in E$ with respect to the $E_i : x = \sum x_i$. The factorization of $\exp(tA)$ we have just given and the fact that, if $k \neq i$, $A_k(x_i) = 0$ and so $\exp(tA_k)x_i = x_i$, imply that $\exp(tA)x = \sum_i \exp(tA_i)x_i$.

Let us write $\exp(tA_i)$ in the form $\exp[t(A_i - \lambda_i.1) + t\lambda_i.1]$. Since 1 and $A - \lambda_i.1$ commute this may also be written as $\exp(tA_i) = \exp(t\lambda_i)$ $\exp[t(A_i - \lambda_i.1)]$. But $A - \lambda_i.1$ is nilpotent, with index r_i; by §5.5.2, $\exp[t(A_i - \lambda_i.1)]$ is a polynomial in t with coefficients in $\operatorname{End}(E)$ and of degree less than r_i. Let us write this as $P_i(t)$. Thus $\exp(tA)x = \sum_i e^{t\lambda_i} P_i(t)x_i$. We have therefore shown that the general solution of $x' = Ax$ is the sum of m vectors of the form $e^{t\lambda_i} X_i(t)$, where $X_i(t)$ is a vector with components which are polynomials in t of degree less than the multiplicity r_i of the eigenvalue λ_i.

Corollary 5.2

If $a_1, \ldots, a_m$ are distinct complex numbers and $p_1(t), \ldots, p_m(t)$ are polynomials in t, then $p_1 = \ldots = p_m = 0$ if $e^{ta_1} p_1(t) + \ldots + e^{ta_m} p_m(t) = 0$ for all $t \in \mathbf{R}$.

A practical method of solution consists of finding the eigenvalues λ_i with their multiplicities r_i, then looking for the polynomials $X_i(t)$ by the method of indeterminate coefficients. To do this we take $X_i(t)$ to be the most general polynomial of degree $r_i - 1$ and with values in E, and write down the condition that $e^{t\lambda_i}X_i(t)$ satisfies $x' = Ax$. After division by $e^{t\lambda_i}$ of the equations thus obtained, we have polynomial equations in t each one of whose coefficients can be equal to zero, for these equations are identities. We thus obtain a system of linear equations which enable us to find those of the coefficients of X_i which cannot remain arbitrary (*a priori* r_i of them are arbitrary since dim $E_i = r_i$).

5.5.4 General case on the field of real numbers

Here E is a real n-dimensional vector space and $A \in \mathrm{End}(E)$.

Since a polynomial of degree n does not necessarily have n real roots, the argument of §5.5.3 does not apply. We shall reduce the matter to the complex case.

We begin by constructing a complex vector space from E. Define on the product $E \times E$ the operations of addition and multiplication by a complex number $a + ib$ by the rules:

$$(x, y) + (x', y') = (x + x', \ y + y'), (a + ib)(x, y) = (ax - by, bx + ay)$$

We obtain a complex vector space E^c, of dimension n, which is called the complexification of E. The subspace $\{(x, 0) : x \in E\}$ is canonically isomorphic to E. We identify this with E and call it the space of real vectors. This enables us to write x in place of $(x, 0)$, and since $(x, y) = (x, 0) + i(y, 0)$, we shall write $x + iy$ in place of (x, y).

Now we define the complexification of $A \in \mathrm{End}(E)$. This is the endomorphism A^c of E^c defined by $A^c(x + iy) = A(x) + iA(y)$. Notice that this leaves the real subspace E invariant.

We shall call the equation $z' = A^c z$ the complexification of the differential equation $x' = Ax$. We easily verify the following properties of its solutions.

(a) A solution $t \rightsquigarrow z(t) \in E^c$ of the complexified equation such that $z(0) \in E$ is real; that is, $z(t) \in E$ for all t. Moreover, this solution is also a solution of the real equation.

(b) The function $t \rightsquigarrow z(t) = x(t) + iy(t)$ is a solution of the complexified equation if, and only if, its real part $x(t)$ and its imaginary part $y(t)$ satisfy the real equation.

We now come to the solution of the equation $x' = Ax$ in E with initial condition $x(0) = x_0 \in E$. Let us solve the complexified equation $z' = A^c z$, with the same initial condition $z(0) = x_0 \in E$, by the method of §5.5.3. By property (a) above the solution will be real and will satisfy $x' = Ax$. This is therefore the desired solution. It remains to express it in a real form, for it is the sum of vectors of the form $e^{\lambda t}X(t)$, where $X(t)$ is a polynomial in t and the eigenvalue λ of A is not necessarily real. By property (b) above the real and imaginary parts of $e^{\lambda t}X(t)$ are solutions of $x' = Ax$. If $\lambda = a + ib$, $a, b \in \mathbf{R}$, we shall thus obtain the solution of $x' = Ax$ by writing it as the sum of terms of the form $e^{at}\cos(bt)X(t)$ and $e^{at}\sin(bt)X(t)$, where $X(t)$ is a polynomial in t with real coefficients, of degree less than the multiplicity of λ.

We next look at some applications of these results.

5.6 HOMOGENEOUS LINEAR DIFFERENTIAL EQUATIONS OF ORDER n WITH CONSTANT COEFFICIENTS

We begin with an example drawn from mechanics.

5.6.1 The harmonic oscillator

A point of mass $m = 1$ moves on a line $\mathbf{R}$, attracted to the origin 0 by a force proportional to its distance from 0 (by means of a string, for example). If t denotes time, the abscissa of the point is a function $t \mapsto q(t)$ which obeys Newton's law: $q'' = -\omega^2 q$, where ω is a positive constant (which depends on the string).

The equation $q'' + \omega^2 q = 0$ contains the second derivative of q. We shall reduce it to a first-order equation by a device: *reduction to first order*. Put $q' = mq' = p$ (the momentum of the point). The preceding equation is equivalent to the first-order linear differential system:

$$q' = p, p' = -\omega^2 q. \tag{5.3}$$

That is, to every solution q of the equation corresponds a solution $(q, p = q')$ of equation (5.3); and conversely, given any solution (q, p) of equation (5.3), q is a solution of the equation $q'' + \omega^2 q = 0$.

The general solution of equation (5.3) is easily obtained:

$$q(t) = a\cos\omega t + b\sin\omega t$$

$$p(t) = \omega(-a\sin\omega t + b\cos\omega t)$$

where the initial position $q(0) = a$ and the initial momentum $p(0) = \omega b$ are arbitrary.

The space $\mathbf{R}^2$ of the points (q, p) is called the phase space. Its interest lies in the fact that through an arbitrary point $(q(0), p(0))$ passes one and only one solution curve (also called an integral curve). These integral curves are of two types: the origin $(0, 0)$, and ellipses with centre 0 and equation $q^2 + \omega^{-2}p^2 = a^2 + b^2$.

Notice also that all the solutions have period $2\pi/\omega$.

Remark

Suppose we have n harmonic oscillators. The phase space is:

$$\mathbf{R}^{2n} = \{q = (q_1, \ldots, q_n), p = (p_1, \ldots, p_n)\}$$

where q_k (resp. p_k) is the position (resp. the momentum) of the kth point mass. The evolution of the mechanical system is determined by the homogeneous linear differential system with constant coefficients $q'_k = p_k$, $p'_k = -\omega_k^2 q_k$, $k = 1, \ldots, n$.

The integral curves $t \rightsquigarrow (q(t), p(t))$ are still bounded, but they are only periodic if, and only if, the partial periods $2\pi/\omega_k$ are commensurable; that is, if there are non-zero integers $c_1, \ldots, c_n$ such that $c_1\omega_1 + \ldots + c_n\omega_n = 0$.

5.6.2 The general case

We say that an n-times differentiable function y: $\mathbf{R} \to \mathbf{C}$ is a solution of a homogeneous linear nth-order differential equation with constant coefficients $a_1, \ldots, a_n$ if it satisfies:

$$y^{(n)} + a_1 y^{(n-1)} + \ldots + a_n y = 0 \tag{5.4}$$

where $y^{(k)}$ is the kth derivative of y.

To solve this equation means finding all the functions y which satisfy it. To do this, let us use the method of reduction to the first order. Set $y = x_l$, $y' = x_2, \ldots, y^{(n-1)} = x_n$; clearly $x = (x_1, \ldots, x_n)$ satisfies the first-order homogeneous linear differential system with constant coefficients:

$$\begin{aligned} x'_1 &= x_2 \\ &\vdots \\ x'_{n-1} &= x_n \\ x'_n &= -a_n x_1 - a_{n-1}x_2 - \ldots - a_1 x_n \end{aligned} \tag{5.5}$$

Conversely, it is plain that the first component x_1 of a solution of equation (5.5) is a solution of equation (6.4). With:

$$A = \begin{pmatrix} 0 & 1 & 0 & \ldots & 0 \\ 0 & 0 & 1 & \ldots & 0 \\ \vdots & & & & 1 \\ -a_n & -a_{n-1} & & \ldots & -a_1 \end{pmatrix}$$

equation (5.5) may be written as $x' = Ax$. The characteristic equation $\det(A - k.1) = 0$ is simply $k^n + a_1 k^{n-1} + \ldots + a_n = 0$, as we can easily see by induction on n. By the results of §5.5.3, if k_i is a root of order r_i of this polynomial, the general solution of $x' = Ax$ is $\sum_1^m \exp(k_i t) P_i(t)$, where $P_i(t)$ is a polynomial of degree $r_i - 1$. Looking merely at the first component, we see that the general solution of equation (5.4) is of the same form, $P_i(t)$ being a polynomial with scalar values.

It may be inconvenient to reduce the integration of equation (5.4) to that of a first-order system. We shall now look at another method, whose range of application considerably extends the present situation.

5.6.3 Differential polynomials with constant coefficients

We return to equation (5.4). A solution y is n times differentiable to begin with. In fact, since $y^{(n)} = -a_1 y^{(n-1)} - \dots - a_n y$, $y^{(n)}$ is differentiable, and so y is $(n+1)$ times differentiable. Lastly, step by step, we see that y is C^∞: every solution of equation (5.4) belongs to $C^\infty = C^\infty(\mathbf{R}; \mathbf{C})$.

Denote by D the linear (discontinuous!) map which takes each $f \in C^\infty$ into its derivative f' and, more generally by $D^k f$ the kth derivative of f: we may regard D^k as the kth power of the operator D. Equation (5.4) may be written, with this notation:

$$D^n y + a_1 D^{n-1} y + \dots + a_n y = 0.$$

Its solution is reduced to the following problem: find the kernel of the operator $p(D) = D^n + a_1 D^{n-1} + \dots + a_n$ in the space C^∞.

We say that $p(D)$ is a differential polynomial with constant coefficients. It is clear that the set of these polynomials forms a ring isomorphic to that of the polynomials in $x \in C$. In particular, if $p(x)$ may be factorized as $p(x) = q(x)d(x)$, we also have $p(D) = q(D)d(D)$. It follows that $y \in \ker p(D)$ if, and only if, $d(D)y \in \ker q(D)$.

We return to equation (5.4), written in the form $p(D)y = 0$ and considered in the context of the field of complex numbers. The polynomial p may be factorized completely and, if k is one of its roots, $p(D) = (D-k)q(D)$. Thus $q(D)y \in \ker(D-k)$. Now an element u of $\ker(D- \;\; k)$ satisfies $u' = ku$, and so $u(t) = A \exp(kt)$, where A is an arbitrary constant. Thus equation (5.4) may be written as $q(D)y = A \exp(kt)$. We may continue by factorizing q. Each step introduces an arbitrary constant. We finish with the term arising from degree (p) steps. Hence the vector space of solutions of $p(D)f = 0$ is of dimension $n = \text{degree } (p)$

We now look for a basis of this space. We shall show that if k is a multiple root of order r of p, then $t \mapsto \exp(kt)t^s$ is in $\ker(p)$ for $0 \leqslant s < r$. Since $(D-k)^r$ divides $p(D)$, it is enough to prove that $(D-k)^r(\exp(kt)t^s) = 0$. As $(D-k)(\exp(kt)t^s) = s \exp(kt)t^{s-1}$, the result follows by iteration. It follows that if $k_1, \dots, k_m$ are the distinct roots of p, of multiplicities $r_1, \dots, r_m$ respectively, then:

$$\exp(k_1 t), \dots, \exp(k_1 t)t^{r_1 - 1}, \dots, \exp(k_m t)t^{r_m - 1}$$

are $r_1 + \dots + r_m = n$ solutions.

It remains to show that these solutions are linearly independent; that is, that if $p_1, \dots, p_m$ are polynomials of degrees $r_1 - 1, \dots, r_m - 1$ respectively, then:

$$\exp(k_1 t)p_1(t) + \dots + \exp(k_m t)p_m(t) = 0$$

implies that $p_1 = \dots = p_m = 0$. This may be proved by induction on m, assuming it true for $m-1$. We divide both sides by $\exp(k_1 t)$, then differentiate (degree p_1) + 1 times. This gives $\exp(k_2 - k_1)t \,.\, P_2(t) + \dots = 0$, where P_i, if it is not zero, is a polynomial of the same degree as p_i. Since $k_2 - k_1 \neq 0$, etc., the inductive hypothesis implies that $p_2 = \dots = p_m = 0$; thus $p_1 = 0$.

5.7 BOUNDED OR PERIODIC SOLUTIONS OF $x' = Ax$

Let E be a complex vector space of dimension n, and let $A \in \text{End}(E)$. We look for conditions under which all solutions of $x' = Ax$ are bounded.

By §5.5.3 the general solution is a linear combination of vectors of the form $\exp(kt)P(t)$, where P is a polynomial of degree less than the multiplicity of the eigenvalue k of the characteristic polynomial of A. If all the solutions are bounded, then since $e^{at}t^k \to +\infty$ as $t \to +\infty$(resp. $-\infty$) if $a > 0$ (resp. $a < 0$), it follows that the eigenvalues are all purely imaginary: $k = i\omega$, where $\omega \in \mathbf{R}$. Let us return to the vector solution $\exp(kt)P(t)$; under these conditions $\| \exp(kt)P(t) \| = \| P(t) \|$, which can remain bounded only if P is of degree zero. The operator A is thus diagonalizable and its matrix in a convenient basis is $\text{diag}(i\omega_1, \ldots, i\omega_n)$, where $\omega_k \in \mathbf{R}$.

Conversely, if A is of this form, then the general solution of $x' = Ax$, written in terms of the above basis, is $(\exp(i\omega_1 t)x_1, \ldots)$. It thus remains bounded. We have accordingly proved the following result:

Theorem 5.9

In order that all solutions of $x' = Ax$ should be bounded, it is necessary and sufficient that A be diagonalizable and that its eigenvalues be purely imaginary.

Remark

By means of the basis above, E may be identified with C^n. We write each of the n copies of C in the form $\{q_r = ip_r : q_r, p_r \in \mathbf{R}\}$. The space E may also be identified with $\mathbf{R}^{2n} = \{(q_1, p_1, \ldots, q_n, p_n)\}$, and the general solution of $x' = Ax$ is:

$$q_r = a_r \cos(\omega_r t) - b_r \sin(\omega_r t)$$

$$p_r = a_r \sin(\omega_r t) + b_r \cos(\omega_r t)$$

This is (see §5.6.1) the general solution of a system of n harmonic oscillators of frequencies ω_r. We deduce the following theorem.

Theorem 5.10

In order that all the solutions of $x' = Ax$ be periodic (hence bounded), it is necessary and sufficient that A be diagonalizable and that its eigenvalues be purely imaginary and commensurable. □

Bibliography

For a more complete account of these questions the reader is referred to the treatise by E. A. Coddington and N. Levinson, and to the book of V. I. Arnold, which is full of examples and written in a lively manner.

6

The integral product, and linear differential equations

This chapter is devoted to linear differential equations with variable coefficients.

PRELIMINARIES

Let A be a continuous endomorphism of a Banach space E. We have seen in Chapter 5 that the general solution of the linear differential equation $x' = Ax$ is $\exp(tA)x_0$, $x_0 \in E$. On the other hand, we known from Definition 5.2 that $\exp(tA) = \lim_{n \to \infty} [1 + (tA/n)]^n$; $[1 + (tA/n)]^n x_0$ is thus an approximate solution of $x' = Ax$; this is Euler's method. We shall generalize it by following a procedure due to V. Volterra (1887).

We no longer assume that A is a constant. If A is a continuous map of an interval I to $\mathrm{End}(E)$, we intend to find a differentiable map $x: I \to E$ such that $x'(t) = A(t)x(t)$ for all $t \in I$, and $x(t_0) = x_0$, where $t_0 \in I$, $x_0 \in E$. This equation may be approximated by a system of finite-difference equations:

$$[x(t_{i+1}) - x(t_i)]/(t_{i+1} - t_i) = A(t_i)x(t_i), i = 0, 1, \dots, n-1$$

where $t_0 < t_1 < \dots < t_n = t$. From these equations we obtain:

$$x(t) = \prod_{i=0}^{n} [1 + A(t_i)(t_{i+1} - t_i)] x_0.$$

We are going to show that this expression converges to the desired solution of $x'(t) = A(t)x(t)$ when the steps $t_{i+1} - t_i$ of the partition tend to zero. In particular, if $t_0 = 0$, $t_{i+1} - t_i = (t - a)/n$ and $A =$ constant, the limit is simply $\exp(tA)x_0$; this is the solution of $x' = Ax$ satisfying $x(0) = x_0$.

6.1 THE INTEGRAL PRODUCT

6.1.1 The integral product of step functions

Let $[a, b]$ be a closed interval, with $a < b$. Divide it into n intervals by points $a = t_0 < t_1 < \dots < t_n = b$; suppose we are given n endomorphisms

$A_1, \ldots, A_n \in \mathrm{End}(E)$. The map $A\colon [a, b] \to \mathrm{End}(E)$, defined by $A(t) = A_1$ for $t_0 \leqslant t < t_1, \ldots, A(t) = A_n$ for $t_{n-1} \leqslant t \leqslant t_n$, is a step-function (see §2.6). Its integral product between a and b is, by definition:

$$P_a^b(A) = \prod_a^b \; [1 + A(t)\,\mathrm{d}t] = \exp(\Delta t_n \,.\, A_n) \ldots \exp(\Delta t_1 \,.\, A_1)$$

where $\Delta t_j = t_j - t_{j-1}$ for $1 \leqslant j \leqslant n$.

The reader should note carefully the order of the factors, since *a priori* $A_1, \ldots, A_n$ do not commute. However, if they do commute, and in particular if A is constant, we have by §5.2.3:

$$P_a^b(A) = \exp(\Delta t_n \,.\, A_n + \ldots + \Delta t_1 \,.\, A_1) = \exp\left[\int_a^b A(t)\,\mathrm{d}t\right].$$

Remark

The integral product is not changed by the addition of points of subdivision to those of the subdivision t_i. For $A(t)$ remains constant between t_i and t_{i+1}; if $t_i < t' \leqslant t_{i+1}$ we then have, by §5.2.3:

$$\begin{aligned}\exp(t_{i+1} - t_i)A &= \exp[(t_{i+1} - t')A + (t' - t_i)A] \\ &= \exp(t_{i+1} - t')A \,.\, \exp(t' - t_i)A.\end{aligned}$$

Lemma 6.1

Let A and B be step functions defined on $[a, b]$ and with values in E. Then:

$$\| P_a^b(A) - P_a^b(B) \| \leqslant (b - a)\mathrm{e}^{(b-a)M} \, \| A - B \|$$

where $\| A \| = \sup \| A(t) \|$ for $a \leqslant t \leqslant b$, and $M = \max(\| A \|, \| B \|)$.

Proof. By the last remark we may suppose that A and B are constant on each of the intervals of the same subdivision $a = t_0 < \ldots < t_n = b$.

Set $a_i = \exp(\Delta t_i \,.\, A_i)$ and $b_i = \exp(\Delta t_i B_i)$. Then:

$$\begin{aligned}\| P(A) - P(B) \| = \| a_n \ldots a_1 - b_n \ldots b_1 \| &\leqslant \| a_n \ldots a_1 - a_n \ldots a_2 b_1 \| \\ &+ \| a_n \ldots a_2 b_1 - a_n \ldots a_3 b_2 b_1 \| + \ldots \\ &+ \| a_n b_{n-1} \ldots b_1 - b_n b_{n-1} \ldots b_1 \| \\ &\leqslant \| a_1 - b_1 \| \, \| a_n \| \ldots \| a_2 \| + \| a_2 - b_2 \| \, \| a_n \| \ldots \\ &\qquad \| a_3 \| \, \| b_1 \| + \ldots \\ &+ \| a_n - b_n \| \, \| b_{n-1} \| \ldots \| b_1 \|.\end{aligned}$$

Let us use the estimate in §5.2.1 and $\| \exp L \| \leqslant \exp \| L \|$ to majorize this last expression. We find that:

$\| P(A) - P(B) \| \leqslant \Delta t_1 \| A_1 - B_1 \| \exp(\Delta t_1 . M) . \exp(\Delta t_n . M) \ldots \exp(\Delta t_2 . M) + \ldots + \Delta t_n \| A_n - B_n \| \exp(\Delta t_n . M) . \exp(\Delta t_{n-1} . M) \ldots \exp(\Delta t_1 . M) \leqslant (b - a) \| A - B \| \exp(b - a) M.$ □

6.1.2 The integral product of regulated functions

Let A: $[a, b] \to \mathrm{End}(E)$ be a regulated function; that is the uniform limit on $[a, b]$ of a sequence A_k of step functions (§2.6.2). By Lemma 6.1:

$$\| P_a^b(A_p) - P_a^b(A_q) \| \leqslant (b-a) e^{(b-a)M} \| A_p - A_q \|$$

where M is a bound for the $\| A_k \|$.

It follows that $P(A_k)$ is a Cauchy sequence in End(E). Since E is a Banach space, End(E) is complete and $P(A_k)$ converges. We see immediately that its limit does not depend on the sequence A_k used to approximate A. It is thus legitimate to define what we shall call the integral product of A between a and b by:

$$P_a^b(A) = \prod_a^b [1 + A(t)\,\mathrm{d}t] = \lim_{k \to \infty} P_a^b(A_k).$$

6.1.3 Properties of the integral product

(a) If $\| A \| = \sup \| A(t) \|$ for $a \leqslant t \leqslant b$, the inequality $\| \exp(L) \| \leqslant \exp \| L \|$ immediately shows that $\| P_a^b(A) \| \leqslant \exp(b-a) \| A \|$.

(b) Lemma 6.1 extends by continuity to all regulated functions A and B.

(c) $P_a^b(A)$ is invertible. The relation $[\exp(L)]^{-1} = \exp(-L)$ shows this directly for a step function:

$$\begin{aligned}[P_a^b(A)]^{-1} &= [\exp(\Delta t_n \cdot A_n) \ldots \exp(\Delta t_1 . A_1)]^{-1} \\ &= \exp(-\Delta t_1 . A_1) \ldots \exp(-\Delta t_n . A_n)\end{aligned}$$

(Note the order of the factors.) This property extends by continuity to any regulated function.

(d) Chasles' relation. Plainly $P_a^b(A) = 1$ $(= \mathrm{id}_E)$.

If $a > b$, the last property enables us to define $P_a^b(A)$ by $[P_b^a(A)]^{-1}$. With this convention, if A: $I \to \mathrm{End}(E)$ is a regulated function defined on an interval I, we have the analogue of Chasles' relation:

$$P_a^c(A) = P_b^c(A) P_a^b(A) \quad \text{for} \quad a, b, c \in I.$$

(Note the order of the factors.) This may easily be verified for a step function, and the general case follows by continuity.

(e) The reader who is curious about the notation $\prod_a^b [1 + A(t)\,\mathrm{d}t]$ may prove, in the same way as in Definition 5.2, that:

$$P_a^b(A) = \lim \prod_{i=0}^{n} [1 + \Delta t_i . A(t_i)]$$

when the greatest of the steps $\Delta t_i = t_{i+1} - t_i$ of the partition $a = t_0 < \ldots < t_n = b$ tends to zero.

Lastly, here is the analogue of Theorem 2.9 which says that a continuous function is the derivative of any of its primitives.

Theorem 6.1

If $a < t < b$ and A: $[a, b] \to \text{End}(E)$ is continuous, then $f(t) = P_a^t(A)$ is differentiable and $f'(t) = A(t)f(t)$.

Proof. Take h so small that $a < t + h < b$. The integral product between t and $t + h$ of the constant endomorphism $A(t)$ is clearly:

$$P_t^{t+h}[A(t)] = \exp[hA(t)] = 1 + hA(t) + o(h).$$

On the other hand, Lemma 6.1 implies that:

$$\| P_t^{t+h}(A) - P_t^{t+h}[A(t)] \| \leqslant |h| \exp(|h| \; \|A\|). \sup_{t \leqslant s \leqslant t+h} \|A(s) - A(t)\|$$

which is $o(h)$ since A is continuous. Hence $P_t^{t+h}(A) = 1 + hA(t) + o(h)$. The theorem now follows from Chasles' relation:

$$P_a^{t+h}(A) - P_a^t(A) = [P_t^{t+h}(A) - 1] P_a^t(A). \qquad \square$$

6.2 HOMOGENEOUS LINEAR DIFFERENTIAL EQUATIONS

Definition 6.1

Let I be an interval in **R** (possibly **R** itself) and let A: $I \to \text{End}(E)$. A differentiable map f: $I \to E$ is said to be a solution of the homogeneous linear first-order differential equation $x'(t) = A(t)x(t)$ if $f'(t) = A(t)f(t)$ for all $t \in I$.

Remarks

(a) If A is constant, we again have the constant coefficient differential equations of the last chapter.

(b) Just as in Chapter 5, it follows that the set S_I of all solutions f: $I \to E$ is a vector space. We shall thus have solved the equation—that is, found all its solutions—if we can exhibit a basis of S_I.

(c) If A is of class C^k, $k \geqslant 0$, then every solution is of class C^{k+1}, as we see immediately by induction on k.

6.2.1 Fundamental theorem

Theorem 6.2

Let E be a Banach space, $x_0 \in E$, I an interval in $\mathbf{R}$, $t_0 \in I$, and let $A: I \to \mathrm{End}(E)$ be continuous. Then there is a unique differentiable map f: $I \to E$ such that:

$$f'(t) = A(t)f(t) \text{ for all } t \in I,$$

$$f(t_0) = x_0.$$

This map is of class C^1 and is given by $f(t) = P_{t_0}^t(A)x_0$.

Proof. The existence of a solution follows immediately from Theorem 6.1.

Let us show that the solution given by the theorem is unique. If g is a second candidate, put $h(t) = \phi(t)^{-1}g(t)$, where $\phi(t) = P_{t_0}^t(A)$. This is differentiable and, by Theorem 4.5:

$$h'(t) = -\phi(t)^{-1}\phi'(t)\phi(t)^{-1}g(t) + \phi(t)^{-1}g'(t).$$

But $\phi'(t) = A(t)\phi(t)$ and $g'(t) = A(t)g(t)$, so that $h'(t) = 0$. By Theorem 2.4, h is constant, and since $\phi(t_0) = 1$ we have $h(t_0) = x_0$. Thus $h(t) = x_0$ and so:

$$g(t) = \phi(t)x_0 = f(t).$$ □

Corollary 6.1

Just as in §5.4.1 (Remarks), we see, on taking $I = \mathbf{R}$, that $x'(t) = A(t)x(t)$ has a maximal solution f: $\mathbf{R} \to E$ with $f(t_0) = x_0$ and such that every other solution g: $I \to E$ satisfying $g(t_0) = x_0$ is the restriction of f. □

Corollary 6.2

Just as in §5.4.1 (Remarks) we see that the vector space S of maximal solutions f: $\mathbf{R} \to E$ of $x'(t) = A(t)x(t)$ is isomorphic to E. □

If E is finite-dimensional, a basis of S (often called a fundamental set of solutions) has dim E elements.

The general solution $P_{t_0}^t(A)$ of $x'(t) = A(t)x(t)$ depends linearly on the initial condition x_0. We shall see how it depends on A.

Comparison Theorem 6.3

Let E be a Banach space, $x_0 \in E$, I an interval, $t_0 \in I$, and let A: $I \to \mathrm{End}(E)$ and B: $I \to \mathrm{End}(E)$ be continuous. Let f: $I \to E$ be the unique solution of $x'(t) = A(t)x(t)$ such that $f(t_0) = x_0$, and let g: $I \to E$ be the unique solution of $x'(t) = B(t)x(t)$ satisfying $g(t_0) = x_0$. Then:

$$\|f(t) - g(t)\| \leqslant e^{|t-t_0|M}|t - t_0|\, \|A - B\|\, \|x_0\|$$

where $\| A - B \|$ is the supremum of $\| A(s) - B(s) \|$ on (t_0, t), and M is an upper bound for $\|A\|$ and $\| B \|$.

Proof. The fundamental theorem gives expressions for f and g. It is now enough to apply Lemma 6.1. □

Remarks

(a) We shall approximate the solution f of $x'(t) = A(t)x(t)$ which satisfies $f(t_0) = x_0$. To do this approximate the continuous function A by a step function: given any $\varepsilon > 0$, uniform continuity ensures the existence of a partition $t_0 < t_1 < \ldots < t_n = t$ such that $\| A - B \| < \varepsilon$ if B: $[t_0, t] \to \text{End}(E)$ is a step function defined by $B(t) = A(t_0)$ for $t_0 \leqslant t \leqslant t_1, \ldots,$ $B(t) = A(t_{n-1})$ for $t_{n-1} \leqslant t \leqslant t_n$. Repetition of the proof of Theorem 6.3 shows that $\exp[\Delta t_n . A(t_{n-1})] \ldots \exp[\Delta t_1 . A(t_0)] x_0$ approximates f to within $\varepsilon . e^{|t - t_0| \|A\|} |t - t_0| \| x_0 \|$.

We often call $P_a^t(A)$ the resolvent, or the resolvent kernel, of the equation $x'(t) = A(t)x(t)$ and denote it by $R(a, t)$.

The preceding work gives no practical means of determining the resolvent explicitly. Nevertheless, we shall now examine a theoretically explicit method.

6.2.2 Dyson's exponential

Let A: $I \to \text{End}(E)$ be continuous. Let $[a, b]$, $a < b$, be contained in the interval I, and let M be an upper bound for $\| A(t) \|$ in $a \leqslant t \leqslant b$.

If c, $t \in [a, b]$, inductively define maps $R_n(c, (t)$: $[a, b] \to \text{End}(E)$:

$$\begin{aligned} R_0(c, t) &= 1 \; (= \text{id}_E) \\ R_{n+1}(c, t) &= \int_c^t A(s) R_n(c, s) \, ds \quad \text{for} \quad n \geqslant 0. \end{aligned} \tag{6.1}$$

We show that $\| R_n(c, t) \| \leqslant (|t - c|^n M^n / n!)$. This is clear for $n = 0$. Suppose it has been proved for n; equation (6.1) then gives:

$$\| R_{n+1}(c, t) \| \leqslant \left| \int_c^t M(|s - c|^n / n!) M^n \, ds \right| \leqslant (|t - c|^{n+1} / (n+1)!) M^{n+1}$$

and the property is established.

The sequence $R_0(c, t) + \ldots + R_n(c, t)$ thus converges uniformly on $[a, b]$ to a limit $S(c, t) = \sum_0^\infty R_n(c, t)$, and equation (6.1) shows that:

$$S(c, t) = 1 + \int_c^t A(s) S(c, s) \, ds$$

It follows that $f(t) = S(c, t)$ is differentiable and that it satisfies the differential equation $f'(t) = A(t) f(t)$, with the initial condition $f(c) = 1$. By Theorem 6.1

and uniqueness, we thus have $f(t) = P_c^t(A)$. Hence, $P_c^t(A) = \sum_0^\infty R_n(c,t)$, where the R_n are determined successively by means of equation (6.1):

$$R_n(c,t) = \int_{c \leqslant s_1 \leqslant \ldots \leqslant s_n \leqslant t} \cdots \int A(s_n) \ldots A(s_1)\, ds_1 \ldots ds_n, \text{ if } c < t$$

(Note the order of the factors.)

6.2.3. Expression for the resolvent by means of a fundamental system of solutions

Choose a basis in the space E, assumed to be of finite dimension n. Let $x_1(t), \ldots, x_n(t)$ be a fundamental system of solutions of the differential equation $x'(t) = A(t)x(t)$. Denote by $X(t)$ the $n \times n$ matrix whose kth column, read from top to bottom, is made up of the components of $x_k(t)$. Since $x_k(t) = P_n^t(A)x_k(a)$, it follows that $X(t) = R(a,t)X(a)$.

Since the vectors $x_1(a), \ldots, x_n(a)$ are linearly independent, the matrix $X(a)$ is invertible, and so $R(a,t) = X(t)X(a)^{-1}$.

Although it was the resolvent which was used to prove the existence of solutions of $x'(t) = A(t)x(t)$, it is this formula which will determine it explicitly if we are lucky enough to know n linearly independent solutions.

In the absence of an explicit expression for the resolvent $R(a,t) = P_a^t(A)$, we may still see how properties of $A(t)$ are reflected in $R(a,t)$. Here shall now look at some important examples.

Theorem 6.4

If A is constant, $R(a,t) = \exp(t-a)A$.

Proof. This is immediate, by §6.1.1, and not surprising. For $\exp(t-a)Ax_0$ is the solution of the constant-coefficient differential equation $x'(t) = Ax(t)$ which satisfies the initial condition $x(a) = x_0$ (see Theorem 5.8). □

Theorem 6.5

In order that $C \in \text{End}(E)$ should commute with $R(a,t)$ for all a, t, it is necessary and sufficient that C should commute with $A(t)$ for all t.

Proof. Suppose that $CR(a,t) = R(a,t)C$ for all a, t. Since:

$$\frac{d}{dt}R(a,t) = A(t)R(a,t)$$

it follows that $CA(t)R(a,t) = A(t)R(a,t)C$; and as $R(a,a) = 1$, then $CA(a) = A(a)C$ for all a.

Conversely, assume that $CA(t) = A(t)C$ for all t. Set $f(t) = CR(a,t) - R(a,t)C$. Since $(d/dt)R(a,t) = A(t)R(a,t)$, we have:

$$f'(t) = CA(t)R(a,t) - A(t)R(a,t)C = A(t)[CR(a,t) - R(a,t)C] = A(t)f(t).$$

Thus f is the solution, with values in End(E), of $x'(t) = A(t)x(t)$ which satisfies $f(a) = CR(a, a) - CR(a, a) = 0$. By uniqueness, $f(t) = 0$ for all t, and hence $CR(a, t) = R(a, t)C$. □

Example

Let the Banach space be the complexification $E^c = E + iE$ of a real space E, and for each t let $A(t)^c$ be the complexification of $A(t) \in \mathrm{End}(E)$ (see §5.5.4). Then for all t, $A(t)^c$ commutes with the canonical projection of E^c on the real space E. It follows from Theorem 6.5 that the solution of $z'(t) = A(t)^c z(t)$ which satisfies $z(t_0) = z_0 \in E$ remains real for all t.

Theorem 6.6

Let b be a continuous bilinear form on E. In order that:

$$b[R(a,t)x, R(a,t)y] = b(x, y) \quad \text{for} \quad x, y \in E \text{ and all } a, t \tag{6.2}$$

it is necessary and sufficient that:

$$b(A(t)x, y) + b(x, A(t)y) = 0 \quad \text{for} \quad x, y \in E \text{ and all } t. \tag{6.3}$$

Proof. Write $f(t)$ for the left-hand side of equation (6.2). By Leibniz' rule:

$$f'(t) = b[A(t)R(a,t)x, R(a,t)y] + b[R(a,t)x, A(t)R(a,t)y].$$

If $f(t) = b(x, y)$, we have $f'(a) = 0$, from which equation (6.3) follows. Conversely if (6.3) holds, $f'(t) = 0$ and $f(t)$ reduces to the constant $f(a) = b(x, y)$. □

Examples

(a) If E is a real (resp. complex) Hilbert space, $R(a, t)$ is orthogonal (resp. unitary) for all a, t if, and only if, the adjoint A^* of A satisfies $A^*(t) = -A(t)$ for all t.

In this case, the scalar (resp. hermitian) product of two solutions of $x'(t) = A(t)x(t)$ does not depend on t.

(b) If E is a real symplectic space (that is, endowed with a non-degenerate bilinear alternating form ω), $A(a, t)$ is symplectic for all a, t if, and only if:

$$\omega[A(t)x, y] + \omega[x, A(t)y] = 0$$

for $x, y \in E$ and all t. We then say that A is an infinitesimal symplectic map.

6.2.4 The Jacobi–Liouville theorem

Theorem 6.7

Suppose that E is of finite dimension n, and let $A: I \to \mathrm{End}(E)$ be regulated.

Then:

$$\det P_a^t(A) = \exp \int_a^t \operatorname{tr}[A(s)]\, ds \quad \text{for} \quad a, t \in I.$$

Proof. If A is a step function, then with the notation of §6.1.1 we have:

$$P_a^t(A) = \exp(\Delta t_n . A_n) \ldots \exp(\Delta t_1 . A_1).$$

The formula results from Theorem 5.5 and from the fact that the determinant of a product is the product of the determinants of the factors.

The general case follows from this. We take a sequence of step functions which converges uniformly on $[a, t]$ to A, and note that the determinant and the trace are continuous functions. □

Example

Recall that a volume v of E is a non-zero n-linear antisymmetric form and that the inverse image $f * v = \det(f)v$. In order that f should preserve a volume v, that is, $f * v = v$, it is thus necessary and sufficient that $\det(f) = 1$.

By Theorem 6.7 we see that $P_a^t(A)$ preserves a volume v of E for all a, t if, and only if, $\operatorname{tr} A(t) = 0$ for all t.

6.3 LINEAR DIFFERENTIAL EQUATIONS WITH A FORCING TERM

Definition 6.2

Let E be a Banach space, I an interval, and let $A: I \to \operatorname{End}(E)$ and $B: I \to E$ be continuous. A differentiable map $f: I \to E$ is said to satisfy the first-order linear differential equation:

$$x'(t) = A(t)x(t) + B(t) \tag{6.4}$$

if $f'(t) = A(t)f(t) + B(t)$ for all $t \in I$. The function B is called the 'forcing term' of the equation.

6.3.1 Uniqueness

Theorem 6.8

Given $x_0 \in E$ and $t_0 \in I$, there is at most one solution f of equation (6.4) such that $f(t_0) = x_0$.

Proof. The difference h between two solutions satisfies the homogeneous linear differential equation $h'(t) = A(t)h(t)$ and $h(t_0) = 0$. By uniqueness (Theorem 6.2), $h(t) = 0$ for all $t \in I$. □

6.3.2 Existence

Let $R_0(t_0, t)$ be the resolvent of the associated homogeneous equation

$x'(t) = A(t) \times t$. We shall look for solutions f of equation (6.4) satisfying $f(t_0) = x_0$, in the form $f(t) = R(t_0, t)g(t)$, where $g\colon I \to E$ is an unknown differentiable function. If the forcing term B of (6.4) were zero, g would reduce to the constant x_0. That is why this method, due to Lagrange, is called the method of variation of constants.

Using the fact that $(\mathrm{d}/\mathrm{d}t)R(t_0, t) = A(t)R(t_0, t)$, we obtain:

$$f'(t) = A(t)R(t_0, t)g(t) + R(t_0, t)g'(t) = A(t)f(t) = R(t_0, t)g'(t).$$

Substitution of this in (6.4) gives $R(t_0, t)g'(t) = B(t)$. We know that the resolvent is invertible and that $R(t_0, t)^{-1} = R(t, t_0)$ (see §6.1.3(c)), so that $g'(t) = R(t, t_0)B(t)$. Integrating and using $g(t_0) = R(t_0, t_0)g(t_0) = f(t_0) = x_0$, we obtain $g(t) = x_0 = \int_{t_0}^{t} R(s,\ t_0)B(s)\,\mathrm{d}s$. Lastly, §2.6.4(d) and Chasles' relation (§6.1.3(d)) gives:

$$f(t) = R(t_0, t)x_0 + \int_{t_0}^{t} R(s, t)B(s)\,\mathrm{d}s. \tag{6.5}$$

Notice that the forcing term is the sum of the general solution $R(t_0, t)x_0$ of the homogeneous equation and of the solution $\int_{t_0}^{t} R(s, t)B(s)\,\mathrm{d}s$ of equation (6.4) which vanishes when $t = t_0$. This latter solution depends linearly on the forcing term B.

Since integration of the forcing term of equation (6.5) is not always easy, we often use, in practice, methods which lie outside the scope of this book (the Laplace transform).

If E is finite-dimensional, we may proceed as follows to integrate equation (6.4). Once a fundamental system of solutions $x_1, \ldots, x_n$ of the homogeneous equation has been found, we look for the solution of (6.4) in the form $f(t) = \sum_r f_r(t)x_r(t)$, where the f_r are differentiable scalar-valued functions. Using the fact that f satisfies (6.4), we obtain:

$$\sum f_r'(t)x_r(t) = B(t).$$

This equation determines the f_r', for the $x_r(t)$ form a basis of E. From it the f_r may be found.

6.3.3 An application: comparison of solutions

We now show how equation (6.5) enables the comparison theorem (Theorem 6.3) to be improved by weakening the hypotheses.

Let I be an interval, F a Banach space and $A\colon I \times F \to \mathrm{End}(E)$ a map which is continuous in $t \in I$ and differentiable in $k \in F$.

Consider the homogeneous linear differential equation, depending on a parameter $k \in F$:

$$x'(t) = A(t, k)x(t). \tag{6.6}$$

Let $f(t, k)$ be its solution which satisfies $f(t_0, k) = x_0$, where $t_0 \in I$ and $x_0 \in E$ are given. We examine how it depends on k and, to do this, we calculate its derivative with respect to k at $k = k_0$. Since $D_1 f(t, k) = A(t, k) f(t, k)$ we have, by Schwarz' theorem (Theorem 4.1):

$$D_1 D_2 f(t, k_0) = D_2 D_1 f(t, k_0) = A(t, k_0) D_2 f(t, k_0) + D_2 A(t, k_0) f(t, k_0).$$

This is an equation of the form $x'(t) = A(t, k_0) x(t) + B(t)$, where $x(t) = D_2 f(t, k_0)$ and $B(t) = D_2 A(t, k_0) f(t, k_0)$, with initial condition $x(t_0) = D_2 f(t_0, k_0) = 0$, for $f(t_0, k) = x_0$ for all k. By equation (6.5) we thus have:

$$D_2 f(t, k_0) = \int_{t_0}^{t} R(s, t, k_0) D_2 A(s, k_0) f(s, k_0) \, ds \tag{6.7}$$

where $R(.,.,k_0)$ is the resolvent of equation (6.6).

Replacing $f(t, k_0)$ by its value $R(t_0, t, k_0) x_0$ and using the arbitrariness of x_0, we obtain a similar formula for $D_2 R(t_0, t, k_0)$.

Example

The reader may apply equation (6.7) to the approximate determination of the solution of the harmonic oscillator equation perturbed by a damping term $k(t)$, where k is a small real parameter and ϕ is a continuous function:

$$q' = p, \quad p' = -\omega^2 q - k\phi p.$$

6.4 LINEAR DIFFERENTIAL EQUATIONS OF ORDER n

Definition 6.3

Let $a_1, \ldots, a_n, b$ be continuous functions defined on an interval I and with values in C. We say that an n times differentiable function F: $I \to C$ is a solution of the nth-order differential equation:

$$y^{(n)} + a_1 y^{(n-1)} + \ldots + a_n y = b \tag{6.8}$$

if f and its derivatives $f^{(k)}$ of order $k = 1, \ldots, n$ satisfy:

$$f^{(n)}(t) + a_1(t) f^{(n-1)}(t) + \ldots + a_n(t) f(t) = b(t) \text{ for } t \in I$$

The function b is called the 'forcing term', and the equation obtained from (6.8) by setting b equal to zero is called the associated homogeneous equation.

We shall reduce equation (6.8) to a first-order vector equation by the method of reduction to first order (see §5.6). Setting $y = x_1$, $y' = x_2, \ldots, y^{(n-1)} = x_n$, we obtain:

$$\begin{aligned} x_1' &= x_2 \\ &\vdots \\ x_{n-1}' &= x_n \\ x_n' &= -a_n x_1 - \ldots - a_1 x_n - b. \end{aligned} \tag{6.9}$$

If f is a solution of (6.8), then $(f, f', \ldots, f^{(n-1)})$ is a solution of (6.9). Conversely, the first component f of a solution of (6.9) is a solution of (6.8).

We introduce the vectors $x(t)$ and $B(t)$ with components $(x_1(t), \ldots, x_n(t))$ and $(0, \ldots, 0, b(t))$ respectively, and the matrix:

$$A(t) = \begin{pmatrix} 0 & 1 & \ldots & 0 \\ 0 & 0 & \ldots & 0 \\ \vdots & \vdots & 1 & \\ -a_n(t) & -a_{n-1}(t) & \ldots & -a_1(t) \end{pmatrix} \tag{6.10}$$

Equation (6.9) may also be written as $x'(t) = A(t)x(t) + B(t)$. Using the results of §§6.3.1 and 6.3.2 we deduce the following theorem.

Theorem 6.9

Given any $t_0 \in I$ and $x_0, \ldots, x_0^{(n-1)} \in C$, equation (6.8) has a unique solution $f: I \to C$ such that $f(t_0, \ldots, x_0^{(n-1)}(x_0) = x_0^{(n-1)})$. The general integral of (6.8) is obtained by adding a particular solution of this equation to the general integral of the associated homogeneous equation. □

Theorem 6.9 reduces the solution of equation (6.8) to that of the associated homogeneous equation and to the discovery of a particular solution.

6.4.1 Determination of a particular solution

We shall apply Lagrange's method of variation of constants.

In certain cases we may be able to use the following remark: if the second member b of equation (6.8) is the sum $\sum b_r$ of several functions, and if s_r is a particular solution of the equation obtained by replacing b by b_r in (6.8), then $\sum s_r$ is a particular solution of (6.8).

Here is an even more special case. Suppose we have to solve the equation $y^{(n)} = b$. This seems to require n successive integrations. But, by Theorem 6.9 it is sufficient to find a function whose nth derivative is known and for which $f(t_0), \ldots, f^{(n-1)}(t_0)$ are arbitrary. Taylor's formula with integral remainder (Theorem 4.7) solves the problem:

$$f(t) = f(t_0) + (t - t_0)f'(t_0) + \ldots + [(t - t_0)^{n-1} f^{(n-1)}(t_0)/(n-1)!]$$
$$+ [1/(n-1)!] \int_{t_0}^{t} (t - s)^{n-1} b(s)\, ds.$$

6.4.2 Solution of the homogeneous equation

Unlike the situation in §5.6.2 there is no manageable formula for the general solution, not even when $n = 2$.

However, let us suppose that a particular non-zero solution s of equation (6.8) is known. Set $y = su$ and determine $y^{(k)}$ with the help of Leibniz' formula. The equation will be transformed into an equation of the same type in which the coefficient of u will be zero since s satisfies (6.8). We thus obtain a linear differential equation of order $n-1$ in u'. If v is the general solution of this equation in u', the general solution of (6.8) will be given by $s\int v + Cs$, where C is an arbitrary constant.

6.4.3 The Wronskian

Suppose we know n solutions $f_1, \ldots, f_n$ of the homogeneous equation associated with equation (6.8). From this we obtain n solutions $x_r = (f_r, f_r', \ldots, f_r^{(n-1)})$, $r = 1, \ldots, n$, of the equation $x'(t) = A(t)x(t)$. From $(x_1, \ldots, x_n)$ we form the matrix $X(t)$ introduced in §6.2.3:

$$X(t) = \begin{pmatrix} f_1(t) & \ldots & f_n(t) \\ \vdots & & \\ f_1^{(n-1)}(t) & \ldots & f_n^{(n-1)}(t) \end{pmatrix}$$

The determinant of $X(t)$ is called the Wronskian $w(t)$ of the solutions $f_1, \ldots, f_n$.

If $R(t_0, t)$ is the resolvent matrix of $x'(t) = A(t)x(t)$, we have seen (§6.2.3) that $R(t_0, t)X(t_0) = X(t)$. Thus $\det R(t_0, t) \cdot \det X(t_0) = \det X(t)$. On the other hand we know (§6.2.4) that:

$$\det R(t_0, t) = \exp \int_{t_0}^{t} \operatorname{tr} A(s)\, ds.$$

Since equation (6.10) immediately gives $\operatorname{tr} A(s) = -a_1(s)$, it follows that:

$$w(t) = w(t_0) \exp\left[-\int_{t_0}^{t} a_1(s)\, ds\right].$$

This shows (Liouville) that if $f_1, \ldots, f_n$ is a basis of the space of solutions defined on an interval I, their Wronskian never vanishes on I. Put another way, if $w(t_0) = 0$, then $w(t) = 0$ for all $t \in I$.

6.4.4 Differential polynomials with variable coefficients

Let us try to adapt the method of solution proposed in §5.6.3 for the solution of the associated homogeneous equation (6.8).

We meet the first difficulty in the construction of a space of solutions. If functions $a_1, \ldots, a_n$ are of class C^r, $r \geqslant 0$, at most, then f is of class C^{n+r}, as (6.8) shows immediately by induction on r. But f is not necessarily of class C^{n+r+1} and, *a fortiori*, does not necessarily belong to the space C^∞. The operator $D = d/dt$ can thus act only on a subspace of the space of solutions.

Let us assume that this difficulty has been overcome (for example, if the a_r are of class C^∞). We try to write, in the simple case $n = 2$, the equation $y'' + a_1 y' + a_2 y = 0$ in the form $(D - A_1)(D - A_2)y = 0$, where this time A_1

and A_2 are no longer constants. We obtain two equations to find A_1 and A_2:

$$A_1 + A_2 = -a_1 \quad \text{and} \quad A_1A_2 - A_2' = a_2.$$

It follows that A_2 must be a solution of the equation $A_2' + a_1A_2 + A_2^2 + a_2 = 0$. This is a Riccati equation, that is, an equation of the form $z' + a + bz + cz^2 = 0$; and Liouville has shown that in general it cannot be integrated by quadratures. The study of those particular cases in which it can be integrated by quadratures depends on Galois theory of differential fields and lies outside the scope of this book.

6.4.5 Return to homogeneous equations with constant coefficients

Consider the homogeneous linear constant-coefficient equation:

$$y^{(n)} + a_1 y^{(n-1)} + \ldots + a_n y = 0. \tag{6.11}$$

Suppose that the characteristic equation $p(k) = k^n + a_1 k^{(n-1)} = \ldots + a_n = 0$ has a multiple root c of order r. Then $p(k) = (k-c)^r q(k)$.

Replace p by the polynomial $(k-c)(k-c-u)\ldots[k-c-(r-1)u]q(k)$. This is the characteristic polynomial of an equation of the form:

$$y^{(n)} + a_1(u) y^{(n-1)} + \ldots + a_n(u) y = 0 \tag{6.12}$$

By §5.6.2 this equation has as solutions $e^{ct}, \ldots, e^{[c+(r-1)u]t}$ and hence also their linear combinations e^{ct}, $e^{ct}[(e^{ut}-1)/u]$, $\ldots$, $e^{ct}[(e^{ut}-1)/u]^{r-1}$, which are linearly independent. When u tends to zero, these functions have limits:

$$e^{ct}, e^{ct}t, \ldots, e^{ct}t^{r-1}. \tag{6.13}$$

Let us use the comparison theorem (Theorem 6.3). The matrix (6.10), formed from $a_n(u), \ldots, a_1(u)$, converges to the matrix formed from $a_n, \ldots, a_1$ if $u \to 0$. It follows that the solutions of equation (6.12) converge to those of (6.11). Hence (6.13) gives r linearly independent solutions of (6.11). We have rederived, by a method due to d'Alembert, the result of §5.6.2.

Bibliography

It is impossible to give a full bibliography, even a summary, for linear differential equations of order n with variable coefficients (even if $n = 2!$). However, the reader is referred to the book by Coddington and Levinson which has already been cited. A study of periodic solutions will also be found in Appendix F.

7

Vector fields and differential equations

This chapter is devoted to differential equations, to the existence of solutions and to the study of the way in which these solutions depend on initial conditions and possible parameters.

7.1 VECTOR FIELDS AND AUTONOMOUS DIFFERENTIAL EQUATIONS

7.1.1 Vector fields

A vector field on an open subset U of a Banach space E is a map $X: U \to E$ of class C^r, $r \geqslant 0$.

It is useful to have in mind the following interpretation. A fluid occupies the interior U of a receptacle in ordinary space. To each point u of U attach the velocity vector $X(t, u)$ of the 'particle' of fluid passing through u at time t. (Mechanists represent it by a 'bound' vector of origin u; that is, since ordinary space is an affine space, by an arrow with origin u and extremity $u + X(t, u)$.) If $X(t, u)$ does not depend on t (we say that the fluid is in steady motion), $u \rightsquigarrow X(u)$ is a vector field on U, called the field of velocity vectors.

Suppose that the flow is steady. If $t \rightsquigarrow f(t) \in U$ is the streamline of a particle, its velocity vector $f'(t)$ at time t is also $X(f(t))$. Knowledge of the trajectories of every particle and their time-like description thus determines the velocity field.

It is the inverse problem which is going to interest us: knowing the velocity field of a steady flow, to determine the time-like equation of each particle.

7.1.2 Integral curves

A differentiable map f: $I \to U$ of an open interval I in U, such that $f'(t) = X(f(t))$ for all $t \in I$, is called an integral curve of a vector field X.

Since X is continuous, so is $X \circ f$ and f is *ipso facto* of class C^1.

We also say that f is a solution of the first-order differential equation $x' = X(x)$. To solve this equation is, by definition, to find all the integral curves of X.

Knowing the position x_0 at time t_0 of a fluid particle, and knowing the velocity field, we may hope that the trajectory of the particle will be found to be determined. This reduces to looking for a solution $f: I \to U$ of $x' = X(x)$ such that $f(t_0) = x_0$. This is the Cauchy problem.

7.1.3 Elimination of the time

Let J be an open interval, U an open subset of a Banach space E, and $X: J \times U \to E$ a C^r map, $r \geqslant 0$. For each $t \in J$ the map $u \in U \leadsto X(t, u) \in E$ is a vector field on U. The fluid flow is no longer steady and the velocity field depends on the time.

Define $\bar{X}: J \times U \to R \times E$ by $\bar{X}(t, u) = (1, X(t, u))$ and regard $\bar{X}$ as a vector field in the open subset $J \times U$ of the Banach space $\mathbf{R} \oplus E$ (this is the space-time of the fluid). The integral curves $s \leadsto (u(s), f(s))$ of $\bar{X}$ thus satisfy $\mathrm{d}u/\mathrm{d}s = 1$, $\mathrm{d}f/\mathrm{d}s = X(u(s), f(s))$. Hence $t = u(s) = s + \text{constant}$, and the integral curves of $\bar{X}$ such that $u(0) = 0$ may be written as $t \leadsto (t, f(t))$; it follows that $\mathrm{d}f/\mathrm{d}t = X(t, f(t))$. The projection on E of this integral curve $(t, f(t)) \in \mathbf{R} \times E$ is thus a solution of the differential equation $x' = X(t, x)$. We say that this equation is *non-autonomous*, in contrast to equations of the form $x' = X(x)$, called *autonomous*, where t does not appear explicitly in X.

Conversely, if $t \leadsto f(t)$ is a solution of $x' = X(t, x)$, then $t \leadsto (t, f(t))$ is clearly an integral curve of $\bar{X}$. The study of 'time-dependent' vector field $X(x, t)$ is thus reduced to that of time-independent vector fields.

7.1.4 Elimination of parameters

Let V be an open subset of a Banach space P, U an open subset of a Banach space E, and $X: U \times V \to E$ a map of class C^r, $r \geqslant 0$. For each $v \in V$ the map $u \in U \leadsto X(u, v) \in E$ is a vector field which depends on the parameter v.

Define $\bar{X}: U \times V \to E \times P$ by $\bar{X}(u, v) = (X(u, v), 0)$ and regard $\bar{X}$ as a vector field on the open subset $U \times V$ of the Banach space $E \oplus P$.

The integral curves $t \leadsto (f(t), v(t))$ of $\bar{X}$ thus satisfy $f'(t) = X(f(t), v(t))$, $v'(t) = 0$. Hence $v(t) = v$ (constant) and $f'(t) = X(f(t), v)$. The projection on E of the integral curve $(f(t), v) \in E \oplus P$ of $\bar{X}$ is therefore a solution of the differential equation $x' = X(x, v)$. The converse is clear: if $t \leadsto f(t)$ is a solution of this last equation, then $t \leadsto (f(t), v)$ is an integral curve of $\bar{X}$.

Thus the study of equations depending on a parameter may be reduced to that of autonomous equations $x' = X(x)$.

7.1.5 Reduction to first order

Let E be a Banach space, U an open subset of the Banach space $E \oplus E$, and $F: U \to E$ a C^r map, $r \geqslant 0$. This defines a vector field $(x, y) \in U \leadsto (y, F(x, y)) \in E \oplus E$, which we shall denote by X. An integral curve

$t \mapsto (f(t), g(t)) \in E \oplus E$ of X thus satisfies $f'(t) = g(t)$, $g'(t) = F(f(t), g(t))$. This shows that f' is differentiable and that $f''(t) = F(f(t), f'(t))$. In other words, $t \mapsto f(t)$ is a solution of the second-order differential equation $x'' = F(x, x')$.

The procedure is general; it has already been used in §5.6 and §6.4. For example, an nth order differential equation:

$$\frac{\mathrm{d}^n x}{\mathrm{d}t^n} = F\left(x, \frac{\mathrm{d}x}{\mathrm{d}t}, \ldots, \frac{\mathrm{d}^{n-1}x}{\mathrm{d}t^{n-1}}\right) \text{ where } x \in E,$$

may be reduced to an autonomous equation on an open subset of $E^n = E \oplus \ldots \oplus E$:

$$\frac{\mathrm{d}x_1}{\mathrm{d}t} = x_2, \ldots, \frac{\mathrm{d}x_{n-1}}{\mathrm{d}t} = x_n, \frac{\mathrm{d}x_n}{\mathrm{d}t} = F(x_1, \ldots, x_n).$$

Remark

The reduction to first order is not canonical; we may reduce an nth order system to a first-order one in various ways. For example, the equation $x'' + \omega^2 x = 0$ of the harmonic oscillator (§5.6.1) may be written as $x' = y, y' = -\omega^2 x$; or as $x' = \omega y, y' = -\omega x$. This reduction is fundamental in mechanics, as the following example makes clear.

Example (Hamilton's equations)

Let U be a real-valued C^1 function (a potential) defined on a real, finite-dimensional euclidean vector space E.

Recall (see Chapter 1) that if $\langle , \rangle$ is the scalar product on E, the gradient grad $U(q)$ of U at $q \in E$ is defined by $\mathrm{d}U(q) = \langle \text{grad } U(q), . \rangle$.

Consider Newton's equation $\mathrm{d}^2q/\mathrm{d}t^2 = -\text{grad } U(q)$ which governs the development in the course of time t of a conservative mechanical system, with potential energy U. This reduces to a first-order system $q' = p$, $p' = -\text{grad } U(q)$ on the (phase) space $E \oplus E$ of position q and momentum p.

Following Lagrange we introduce the symplectic (bilinear, alternating and non-degenerate) form ω, defined on $E \oplus E$ by $\omega[(q_1, q_1), (q_2, p_2)] = \langle q_1, p_2 \rangle - \langle q_2, p_1 \rangle$. It defines an isomorphism $a \in (E + E)^* \to X_a \in E \oplus E$ by $a(v) = \omega(X_a, v)$ for $v \in E \oplus E$.

Again following Lagrange we introduce the 'total energy' function $H(p, q) = \frac{1}{2}\langle p, p \rangle + U(q)$ and look for the image of its derivative under the above ismorphism. We easily find that $X_{\mathrm{d}H(q,p)} = (p, -\text{grad } U(q))$. Hence Newton's equation $q'' = -\text{grad } U(q)$ reduces to the first-order equation $m' = X_{\mathrm{d}H(m)}$, where $m = (q, p)$. This is the *hamiltonian form* of the equation of evolution, which in an orthonormal basis of E takes the form familiar to those interested in mechanics: $q_i' = \partial H/\partial p_i$, $p_i' = -\partial H/\partial q_i$, $i = 1, \ldots, n$.

7.2 EXISTENCE AND UNIQUENESS OF INTEGRAL CURVES

7.2.1 Lipschitz maps

Recall that a map $X: U \to F$ of an open subset U of a normed v.s. E to a normed v.s. F is called K-Lipschitz if there is a number $K > 0$ such that $\| X(x) - X(y) \|_F \leqslant K \| x - y \|_E$ for all $x, y \in U$. Clearly X is then continuous.

The mean-value theorem (§2.1) shows that if X is of class C^1, then given any u in U there is a neighbourhood V in U such that the restriction of X to V is lipschitzian. We say that X is locally lipschitzian.

Theorem 7.1

Let U be an open subset of a Banach space E and $X: U \to E$ a K-Lipschitz vector field. Let $x_0 \in U$ and suppose that $r > 0$ is so small that the closed ball $B_r(x_0) = \{x \in E : \| x - x_0 \| \leqslant r\}$ is contained in U. Let M be an upper bound for $\| X(x) \|$ in this ball and set $a = r/M$. Then given any $t_0 \in \mathbf{R}$, there exists a unique differentiable map $f: [t_0 - a, t_0 + a] \to B_r(x_0)$ such that:

$$f'(t) = X(f(t)) \text{ and } f(t_0) = x_0. \tag{7.1}$$

Proof. Given the continuity of f', equations (7.1) amount to

$$f(t) = x_0 + \int_{t_0}^{t} X[f(s)]\, \mathrm{d}s. \tag{7.2}$$

Consider the space of continuous maps of $[t_0 - a, t_0 + a]$ to $B_r(x_0)$, endowed with the uniform convergence metric. The subset G of its elements u such that $u(t_0) = x_0$ is a closed subspace of it; G is thus a complete metric space.

If $u \in G$, define Tu by $Tu(t) = x_0 + \int_{t_0}^{t} X[u(s)]\, \mathrm{d}s$, where $|t - t_0| \leqslant a$. Clearly Tu is continuous and $Tu(t_0) = x_0$. On the other hand, $Tu \in G$ for, by the definitions of M, a and r:

$$\| Tu(t) - x_0 \| \leqslant \left\| \int_{t_0}^{t} X[u(s)]\, \mathrm{d}s \right\| \leqslant M |t - t_0| \leqslant Ma = r.$$

Thus T maps G to G.

Returning to equation (7.2), we see that f satisfies (7.1) if, and only if, f is a fixed point of T. If we can show that an iterate of T is a contraction of G, the theorem will be proved by Corollary B.2 in Appendix B.

Let $u, v \in G$. Then:

$$\| Tu(t) - Tv(t) \| = \left\| \int_{t_0}^{t} [X(u(s)) - X(v(s))]\, \mathrm{d}s \right\|$$

$$\leqslant K \left| \int_{t_0}^{t} \| u(s) - v(s) \|\, \mathrm{d}s \right| \leqslant K |t - t_0|\, \mathrm{d}(u, v).$$

Replace u and v by Tu and Tv respectively in the last estimate:

$$\| T^2u(t) - T^2v(t) \| \leqslant K \left| \int_{t_0}^{t} \| Tu(s) - Tv(s) \| \, ds \right|$$

$$\leqslant K^2 d(u, v) \left| \int_{t_0}^{t} (s - t_0) \, ds \right| \leqslant \tfrac{1}{2} K^2 |t - t_0|^2 \, d(u, v).$$

By induction we see that $\| T^k u(t) - T^k v(t) \| \leqslant K^k (|t - t_0|^k / k!) \, d(u, v)$. In particular, $d(T^k u, T^k v) \leqslant ((Ka)^k / k!) \, d(u, v)$. If k is large enough, $(Ka)^k < k!$ and the map T^k is a contraction. □

Example

Let A be an endomorphism of a Banach space E. We look for f: $\mathbf{R} \to \mathrm{End}(E)$ such that $f'(t) = Af(t)$, $f(0) = 1$. This can be written as:

$$f(t) = 1 + A \int_0^t f(s) \, ds$$

By application of the previous method, we find that the successive approximations $f_0(t) = 1, \ldots, f_n = T^n f_0$ of f may be expressed as $f_n(t) = 1 + tA + \ldots + t^n A^n / n!$. Thus $f(t) = e^{tA}$ and we recover the fact that $(e^{tA})' = Ae^{tA}$.

Corollary 7.1

Let U be an open subset of a Banach space E, X: $U \to E$ a lipschitzian vector field. If $f_1 : I_1 \to U$ and $f_2 : I_2 \to U$ are two solutions of $x' = X(x)$ which satisfy the same initial condition $f_1(t_0) = f_2(t_0) = x_0$, then f_1 and f_2 coincide on $I_1 \cap I_2$.

Proof. This is an immediate consequence of uniqueness. □

7.2.2 Maximal solutions

By definition, a solution f: $I \to U$ of $x' = X(x)$ extends a solution $f_1 : I_1 \to U$ if $I_1 \subset I$ and if $f(t_0) = f_1(t_0)$ for some $t_0 \in I_1$. By Corollary 7.1, f coincides with f_1 on I_1, which justifies the terminology.

By definition, a solution is maximal if it cannot be extended. Let us show that such solutions exist.

Theorem 7.2

Every solution of $x' = X(x)$ is contained in a unique maximal solution.

Proof. Consider the set of all solutions $f_a : I_a \to U$ which extend a given solution $f_0 : I_0 \to U$. The union I of the I_a is an interval, for it is the union of intervals containing I_0. We shall define a solution f: $I \to U$ which, clearly, will be maximal.

If $t \in I$, there exists I_a such that $t \in I_a$. Set $f(t) = f_a(t)$. If t belongs to another interval I_b, since f_a and f_b coincide on I_0 we have $f_a(t) = f_b(t)$ by equation (7.2). Thus $f(t)$ does not depend on f_a. We have thus defined a map $f: I \to U$. It certainly satisfies $x' = X(x)$ for each $t \in I$, since it coincides with a solution f_a on an interval I_a which contains t. □

7.2.3 An application

Theorem 7.2 enables us to check whether a list of solutions is exhaustive. For example, consider the equation $x' = x^2$, $x \in \mathbf{R}$. Each of the solutions $f_0(t) = 0$ for all $t \in \mathbf{R}$, $p_a(t) = (a - t)^{-1}$ for $t > a$, $n_a(t) = (a - t)^{-1}$ for $t < a$ is maximal.

To each initial condition (t_0, x_0) corresponds one and only one of these solutions: for $x_0 = 0$ it is f_0; for $x_0 > 0$ (resp. < 0) it is n_a (resp. p_a), where $a = x_0^{-1} + t_0$.

Since $X(x) = x^2$ is locally lipschitzian §7.2.2 may be applied and we certainly have all the maximal solutions. We observe that these solutions, except for f_0, are not defined on the whole of $\mathbf{R}$. We shall see later what conditions on X should be imposed to avoid this pathology.

Remark

If the vector field X is continuous but not locally lipschitzian, maximal solutions may not be unique. For example the equation $x' = 3(x^2)^{1/3}$ has two distinct maximal solutions $f(t) = 0$ and $g(t) = (t - t_0)^3$ which are zero when $t = t_0$. This is because $X(x) = 3x^{2/3}$ has an 'infinite' derivative at $x = 0$. Nevertheless, there is the following result, due to Arzela, which we shall not prove.

Theorem 7.3

Let U be an open subset of a finite-dimensional Banach space E, $X: U \to E$ a continuous vector field. Then given any $t_0 \in \mathbf{R}$ and any $x_0 \in U$, there is at least one solution $f: I \to U$ of the equation $x' = X(x)$ such that $f(t_0) = x_0$, $t_0 \in I$. □

7.2.4 Generalization for differential equations of arbitrary order

Consider again the nth order differential equation §7.1.5: $x^{(n)} = F(x, x', \ldots, x^{(n-1)})$. Assume that the map F is lipschitzian with respect to the set of variables $x, \ldots, x^{(n-1)}$. Then given any $t_0 \in \mathbf{R}$ and $x_0, x_1, \ldots, x_{n-1} \in E$, there is a unique maximal solution f such that $f(t_0) = x_0, f'(t) = x_1, \ldots, f^{(n-1)}(t_0) = x_{n-1}$

The reader may rephrase for himself Theorem 7.1 in terms of differential equations which are non-autonomous or which depend on a parameter.

7.3 DEPENDENCE ON INITIAL CONDITIONS

Theorem 7.1 solves the Cauchy problem: given $t_0 \in \mathbf{R}$ and $x_0 \in U$, if $a > 0$ is small enough, there is a unique curve X defined on $[t_0 - a, t_0 + a]$ and taking the value x_0 when $t = t_0$. Denote it by $\phi(t, x_0)$. We shall study $x_0 \rightsquigarrow (t, x_0)$.

Lemma 7.1

(Gronwall's lemma) Let u and v be continuous maps of $[a, b]$, $a < b$, to the non-negative reals. Suppose there is a number $A \geqslant 0$ such that:

$$u(t) \leqslant A + \int_a^t u(s)v(s)\,\mathrm{d}s \text{ for } a \leqslant t \leqslant b. \tag{7.3}$$

Then $u(t) \leqslant A \exp[\int_a^t v(s)\,\mathrm{d}s]$ for $a \leqslant t < b$.

Proof. If $h(t)$ denotes the right-hand side of equation (7.3), we have $h'(t) = u(t)v(t)$. Since, by hypothesis, $u(t) \leqslant h(t)$ and $v(t) \geqslant 0$, it follows that $h'(t) \leqslant h(t)v(t)$. Put:

$$C(t) = h(t) \exp\left[-\int_a^t v(s)v\,\mathrm{d}s\right].$$

Then $C'(t) \leqslant 0$, and so $C(t) \leqslant A$ for $a \leqslant t \leqslant b$, for $C(a) = h(0) = A$. By the definition of C this gives $h(t) \leqslant A \exp[\int_a^t v(s)\,\mathrm{d}s]$. The lemma follows as $u(t) \leqslant h(t)$, □

Theorem 7.4 (The lipschitzian case)

Suppose the hypotheses of Theorem 7.1 hold. Then given any $x \in B_{r/2}(x_0)$, there is a unique integral curve $\phi(t, x)$ of $x' = X(x)$ defined on $[t_0 - a/2, t_0 + a/2]$ and satisfying $\phi(t_0, x) = x$.

Moreover $\| \phi(t, x) - \phi(t, y) \| \leqslant \mathrm{e}^{K|t - t_0|} \| x - y \|$ for all $x, y \in B_{r/2}(x_0)$ and $|t - t_0| \leqslant a/2$. In particular, $x \rightsquigarrow \phi(t, x)$ is lipschitztian on $B_{r/2}(x_0)$ uniformly in t:

$$\| \phi(t, x) - \phi(t, y) \| \leqslant \mathrm{e}^{Ka/2} \| x - y \|.$$

Proof. The first part is simply a reformulation of Theorem 7.1. If $x \in B_{r/2}(x_0)$ we replace the point x_0 by the point x and $B_r(x_0)$ by $B_{r/2}(x)$ (which still belongs to the open set U). We also replace the number a in by $a/2$ so that $\phi(t, x)$ lies in $B_{r/2}(x)$.

If $x, y \in B_{r/2}(x_0)$, $\phi(t, x)$ and $\phi(t, y)$ satisfy equation (7.2), with suitable change of notation. Put $u(t) = \| \phi(t, x) - \phi(t, y) \|$. Since X is K-lipschitzian, we have:

$$u(t) = \left\| x - y + \int_{t_0}^t [X(\phi(s, x)) - X(\phi(s, y))]\,\mathrm{d}s \right\|$$

$$\leqslant \| x - y \| + K \left| \int_{t_0}^t \| \phi(s, x) - \phi(s, y) \|\,\mathrm{d}s \right| = \| x - y \| + K \left| \int_{t_0}^t u(s)\,\mathrm{d}s \right|$$

and the theorem follows from Gronwall's lemma if $t \geqslant t_0$.

If $t_0 > t$, we may reduce the matter to the preceding case by changing t to $-t$ and X to $-X$.

Theorem 7.5 (Class C^1)

Suppose the hypotheses of Theorem 7.1 hold, save that we assume that X is of class C^1. We then know (see §2.1) that X is locally lipschitzian. Taking a smaller open set U, we may also assume that X is K-lipschitzian for some K. Then the conclusions of Theorem 7.4 hold. But we shall show, in addition, that ϕ is of class C^1.

Proof. If we knew that ϕ was sufficiently differentiable, Schwarz' theorem (§4.1) and the equation $\mathrm{d}\phi/\mathrm{d}t = X \circ \phi$ would imply that:

$$\frac{\mathrm{d}}{\mathrm{d}t} D_2\phi(t, x) = DX[\phi(t, x)] \circ D_2\phi(t, x)$$

with $D_2\phi(t_0, x) = 1$, since $\phi(t_0, x) = x$.

This leads us to study the solution u of:

$$\frac{\mathrm{d}}{\mathrm{d}t} u(t) = DX[\phi(t, x)]\, u(t) \tag{7.4}$$

such that $u(t_0) = 1$ ($= \mathrm{id}_E$). This is a linear differential equation in u whose coefficient $DX[\phi(\ )]$ is continuous since X is of class C^1. The desired solution thus exists, it is unique and is defined for all t (Theorem 6.2). We denote it by $\psi(t, x)$.

(a) *To show that ψ is continuous.* By Theorem 6.2, the map $t \rightsquigarrow \psi(t, x)$ is continuous and even of class C^1 for all x. If we can show that $x \rightsquigarrow \psi(t, x)$ is uniformly continuous in t, the result will follow.

Since DX and ϕ are continuous, we may choose the radius r of the ball $B_r(x_0)$ and the number $a > 0$ so small that $\| DX[\phi(t, x)] \|$ is bounded above by a number m for all $t \in [t_0 - a/2, t_0 + a/2]$ and all $x \in B_{r/2}(x_0)$.

With this choice of r, since ϕ is continuous, the set $\{\phi(t, x) : |t - t_0| \leqslant a/2, x \in B_{r/2}(x_0)\}$ is compact. The continuous function $\phi(t, x) \rightsquigarrow DX[\phi(t, x)]$ is thus uniformly continuous on this compact set: given any $\varepsilon > 0$ there exists $\delta > 0$ such that:

$$\sup_{|t - t_0| \leqslant a/2} \| \phi(t, x) - \phi(t, y) \| \leqslant \delta \tag{7.5}$$

implies that:

$$\sup_{|t - t_0| \leqslant a/2} \| DX[\phi(t, x)] - DX[\phi(t, y)] \| \leqslant \varepsilon \tag{7.6}$$

By Theorem 7.4, ϕ is lipschitzian in x uniformly in t; thus there exists $\delta' > 0$ such that $\| x - y \| \leqslant \delta'$ implies equation (7.5) and hence (7.6).

Now consider the functions $\psi(t, x)$ and $\psi(t, y)$ defined by

$$\frac{d\psi}{dt}(t, x) = DX[\phi(t, x)]\psi(t, x), \quad \psi(t_0, x) = 1,$$
$$\frac{d\psi}{dt}(t, y) = DX[\phi(t, y)]\psi(t, y), \quad \psi(t_0, y) = 1. \tag{7.7}$$

These are solutions of linear differential equations. By the comparison theorem (Theorem 6.3) we thus have:

$$\begin{aligned}\| \psi(t, x) - \psi(t, y) \| &\leqslant \tfrac{1}{2} a\, e^{am/2} \sup_{|t - t_0| \leqslant a/2} \| DX[\phi(t, x)] - DX[\phi(t, y)] \| \\ &\leqslant a\, e^{am/2} \varepsilon/2\end{aligned}$$

for all $|t - t_0| \leqslant a/2$ if $\| x - y \| \leqslant \delta'$. This is the desired uniform continuity.

(b) *To show that* $D_2\phi(t, x)$ *exists and equals* $\psi(t, x)$. Set $\theta(t, h) = \phi(t, x + h) - \phi(t, x)$. Since $d\phi/dt = X \circ \phi$, and since

$$\theta(t_0, h) = \phi(t_0, x + h) - \phi(t_0, x) = (x + h) - x = h$$

we have:

$$\theta(t, h) = h + \int_{t_0}^{t} \{X[\phi(s, x + h)] - X[\phi(s, x)]\}\, ds.$$

On the other hand, by eqution (7.7) we have:

$$\psi(t, x)h = h + \int_{t0}^{t} DX[\phi(s, x)]\psi(s, h)h\, ds.$$

It follows that:

$$\theta(t, h) - \psi(t, x)h = \int_{t_0}^{t} DX[\phi(s, x)]\,[\theta(s, h) - \psi(s, x)h]\, ds + \int_{t_0}^{t} \{\ \}\, ds \tag{7.8}$$

where $\{\ \} = X[\phi(s, x + h)] - X[\phi(s, x)] - DX[\phi(s, x)]\theta(s, h)$.

We now study $\{\ \}$. Since, by Theorem 7.4, $x \rightsquigarrow \phi(., x)$ is lipschitzian, uniformly in t, there exists $C > 0$ such that $\| \theta(s, h) \| = \| \phi(s, x + h) - \phi(s, x) \| \leqslant C \| h \|$ for $|s - t_0| < a/2$. Since X is of class C^1, given any $\varepsilon > 0$ there exists $\delta > 0$ such that $\| \theta(s, h) \| < \delta$ implies $\| \{\ \} \| \leqslant \varepsilon \| \theta(s, h) \|$. If we choose h so that $C \| h \| < \delta$, we then have $\| \{\ \} \| \leqslant \varepsilon C \| h \|$, and the second term on the right-hand side of equation (7.8) is dominated by $|t - t_0| \varepsilon C \| h \|$, that is, again by $a\varepsilon C \| h \|$.

Apply Gronwall's inequality taking for u the norm of the left-hand side of equation (7.8), for A the number $a\varepsilon C \| h \|$ and for v the majorant m of $\| DX(\,) \|$. We obtain:

$$\| \phi(t, x + h) - \phi(t, x)h - \psi(t, x)h \| \leqslant (\text{constant})\varepsilon \| h \|.$$

It follows that $D_2\phi(t, x) = \psi(t, x)$.

Since ϕ: $(t, x) \mapsto \phi(t, x)$ has continuous partial derivatives $D_1\phi = \mathrm{d}\phi/\mathrm{d}t$ and $D_2\phi$, we see from Theorem 2.5 that ϕ is of class C^1. □

The reader will find in Appendix G a proof which is short and tricky, but conceptually more difficult, due to J. Robbin.

Theorem 7.6

(Class C^k) Suppose the hypotheses of Theorem 7.1 hold, save that we assume that X is of class C^k, $k \geqslant 1$. Then ϕ is of class C^k.

Proof. If $k = 1$ the theorem has just been established. Suppose that $k \geqslant 2$ and assume that the theorem is true up to order $k - 1$. Then ϕ is of class C^1, and from the equation:

$$\frac{\mathrm{d}}{\mathrm{d}t}\phi(t, x) = X[\phi(t, x)]$$

it follows that:

$$\frac{\mathrm{d}}{\mathrm{d}t}\frac{\mathrm{d}}{\mathrm{d}t}\phi(t, x) = DX[\phi(t, x)]X[\phi(t, x)]$$

and

$$\frac{\mathrm{d}}{\mathrm{d}t}D_2\phi(t, x) = DX[\phi(t, x)]D_2\phi(t, x).$$

These three relations form a differential system in the unknown $(\phi, \mathrm{d}\phi/\mathrm{d}t, D_2\phi)$. Since X is of class C^k, the right-hand sides are of class C^{k-1} with respect to this unknown. By the inductive hypothesis, ϕ, $\mathrm{d}\phi/\mathrm{d}t$ and $D_2\phi$ are thus of class C^{k-1}. Hence ϕ is of class C^k.

In particular, if X is C^∞ then ϕ is C^∞.

7.3.1 An application

Theorem 7.7

Let A be an endomorphism of a Banach space E. Then $A \mapsto \exp(A)$ is of class C^∞.

Proof. The proof is due to H. Poincaré. Let us look for the solution of the differential system in $x, y \in \mathrm{End}(E)$; $\mathrm{d}x/\mathrm{d}t = y \circ x$, $\mathrm{d}y/\mathrm{d}t = 0$ satisfying the initial conditions $x(0) = 1$ $(= \mathrm{id}_E)$, $y(0) = A$. We have seen in Chapter 5 that this solution is $x(t) = \exp(tA)$, $y(t) = A$. As the vector field $(x, y) \mapsto (y \circ x, 0)$ is of class C^∞, this solution depends in a C^∞ way on the initial condition A. In particular, $A \mapsto \exp(1 \,.\, A)$ is C^∞. □

We may attempt to find an explicit expression for the derivative of $X \mapsto \exp(X)$. Here, without proof, is what we obtain.

Let $A \in \text{End}(E)$ and denote by $ad(A)$ the map of $\text{End}(E)$ to $\text{End}(E)$ defined by $X \rightsquigarrow A \circ X - X \circ A$. This is a continuous linear map, for:

$$\| ad(A)X \| \leqslant 2 \| A \| \ \| X \|.$$

The derivative is:

$$D(\exp)(A) = e^A \sum_0^\infty [(-1)^{n+1}/(n+1)!] \ [ad(A)]^n.$$

We should be able to argue directly. If $h \in \text{End}(E)$, the series:

$$\sum_1^\infty (A^{n-1}h + A^{n-2}hA + \ldots + hA^{n-1})/n!$$

converges normally on every compact subset of $\text{End}(E)$, for the modulus of its general term is dominated by $\| A \|^{n-1} \| h \|/(n-1)!$. If $L(h)$ denotes its sum, we see that L is linear and that $\| L \| \leqslant e^{\| A \|}$. As we have

$$\| e^{A+h} - e^A - L(h) \| \leqslant \sum_0^\infty [(\| A \| + \| h \|)^n - \| A \|^n - n \| A \|^{n-1} \| h \|]/n!$$

$$= e^{\| A \| + \| h \|} - e^{\| A \|} - \| h \| e^{\| h \|} = o(\| h \|).$$

this shows that 'exp' is differentiable and that $D(\exp)(A)h = L(h)$.

7.4 COMPLETE VECTOR FIELDS

Definition 7.1

Given an open subset of a Banach space E and a locally lipschitzian vector field X on U, we say that X is complete if every maximal integral curve of X is defined on the whole of $\mathbf{R}$; in other words, if for each $t_0 \in \mathbf{R}$ and each $x_0 \in U$, there is a differentiable map $f: \mathbf{R} \to U$ such that $f'(t) = X[f(t)]$ and $f(t_0) = x_0$.

Examples

(a) By Chapters 4 and 5 the maximal solutions of linear differential equations are defined on the whole of $\mathbf{R}$.

(b) On $E = \mathbf{R}$ the constant vector field $X = 1$ is complete. On the other hand, the same vector field, but considered on the open set $U = \mathbf{R}\backslash\{0\}$, is not complete: the general solution $f(t) = t - c$, where c is a constant, is not defined for $t = c$ since f may not take the value zero.

We have already seen that the vector field $X(x) = x^2$ on $\mathbf{R}$ is not complete: the solutions $(c-t)^{-1}$, where c is a constant, are not defined for $t = c$.

In the first example, the field is not complete because it lacks one point in the space $\mathbf{R}\backslash\{0\}$ (which is incomplete!). In the second example, the field is incomplete because $X(x)$ grows too quickly with x.

In mechanics, the trajectories of material points are solutions of differential equations. It is essential that these solutions should exist for all time t. That is why we shall give several criteria for completeness.

Theorem 7.8

Let X be a locally lipschitzian vector field defined on an open subset U of a Banach space E. Let f: $]T_-, T_+[\ \rightarrow U$ be a maximal integral curve of X. Then either $T_+ = +\infty$ or $T_+ < +\infty$ and, in this latter case, given any compact subset K of U there exists $\varepsilon > 0$ such that $f(t) \notin K$ if $t > T_+ - \varepsilon$. In other words, $f(t)$ is eventually outside every compact set.

The corresponding property holds for T_-.

Proof. First notice that, by Theorem 7.2, a maximal integral curve is defined on an interval. Let $]T_-, T_+[$ be this interval.

Suppose that $T_+ < +\infty$, and assume, for the sake of contradiction, that there is no such ε as in the statement of the theorem. Then there is a sequence of real numbers t_n which converges to T_+ and is such that $f(t_n)$ remains in K. Since K is compact, by passage to a subsequence of t_n if necessary, we may assume that $f(t_n)$ converges to a point x_0 of k. By Theorem 7.4, there exists a neighbourhood $U = B_{r/2}(x_0)$ and a number $a > 0$ such that, given any $x \in U$, there is an integral curve $\phi(t, x)$ defined on $[t_0 - a/2, t_0 + a/2]$. Take n so large that $T_+ - t_n < a/2$ and $f(t_n) \in U$. We then have the following contradiction: the maximal integral curve passing through $f(t_n)$ at time t_n is defined only for $t < T_+$, yet we may extend it up to the time $t_n + \frac{1}{2}a > T_+$.

7.4.1 The first integral

The notation of Theorem 7.8 is retained. We say that a function h: $U \rightarrow \mathbf{R}$ is a first integral of the field X if, for all integral curves f: $I \rightarrow U$ of X, the function $h \circ f$ defined on I does not depend on t (but it may depend on the solution f). We shall look at a simple criterion which enables us to decide if a function h is a first integral, without having to know the integral curves explicitly.

Lemma 7.2

A C^1 function h: $U \rightarrow \mathbf{R}$ is a first integral of X if, and only if, $Dh(x)X(x) = 0$ for all $x \in U$.

Proof. Let h be a first integral, and let $x \in U$. If f is an integral curve of X such that $f(a) = x$, we have $h[f(t) = h[f(a)] = h(x)$. Differentiating with respect to t we have:

$$Dh[f(t)]f'(t) = Dh[f(t)]X[f(t)] = 0.$$

Taking $t = a$, we obtain $Dh(x)X(x) = 0$.

Conversely, suppose that $Dh(x)X(x) = 0$ for all x in U. The last calculation shows that the derivative of $t \to h[f(t)]$ is zero. As t runs through a connected interval, it follows from §2.2 that $h[f(t)]$ is constant.

We now return to the completeness of vector fields.

Corollary 7.2

Let X be a locally lipschitzian vector field defined on an open subset U of a Banach space E. If X has a first integral $h: U \to \mathbf{R}$ such that for all $r \in \mathbf{R}$, $h^{-1}(r)$ is compact, then X is complete.

Proof. Each integral curve passing through $x \in U$ remains in the compact set $h^{-1}[h(x)]$. The result thus follows from Theorem 7.8.

Remarks

(a) It may be that the function h is defined on a subset larger than U; on the whole of E, for example. If this is so, the hypothesis means that $h^{-1}(r) \cap U$ must be a compact subset of U.

(b) If h is differentiable, it is easy to see that the hypothesis of the corollary imply that E is finite-dimensional.

Example

The vector field $X = (y - z, z - x, x - y)$ defined on $\mathbf{R}^3 = \{(x, y, z)\}$ has $h(x, y, z) = x^2 + y^2 + z^2$ as a first integral, for $Dh \,.\, X = 2x(y - z) + 2y(z - x) + 2z(x - y) = 0$. Since $h^{-1}(r)$ is empty if $r < 0$ and is the sphere with centre 0 and radius $r^{1/2}$ if $r > 0$, the field X is complete.

7.4.2 A theorem about a *a priori* bounds

Theorem 7.9

Let the hypotheses of Corollary 7.2 hold. Let f: $]T_-, T_+[\to U$ be a maximal integral curve of X. Suppose that given any $T > 0$ there exists a compact subset K_T of U such that $f(t) \in K_T$ for $|t| \leqslant T$. Then $T_- = -\infty$ and $T_+ = +\infty$.

Proof. On the open subset $\mathbf{R} \times U$ of the Banach space $\mathbf{R} \oplus E$ we consider the vector field $(1, X)$; that is, the system $dt/ds = 1$, $dx/dt = X(x)$. Suppose that $T_+ < +\infty$. Take $T > T_+$ and from the compact subset $[-T, +T] \times K_T$ of $\mathbf{R} \times U$. By Theorem 7.8, applied to the field $(1, X)$, the maximal integral curve $]T_-, T_+[\to \mathbf{R} \times U$ defined by $s \mapsto (s, f(s))$ leaves this compact set. As $f(s)$ remains in K_T, this means that there exists s such that $s > T > T_+$: this contradicts the fact that $f(s)$ is not defined for $s \geqslant T_+$.

The proof for T_- is the same.

Thus an *a priori* bound $f(t) \in K_T$ for a solution f on the interval $|t| \leqslant T$ implies the existence of this solution on every interval.

Examples

(a) A locally lipschitzian vector field X defined on all of E and constant ($=X_0$) outside of a compact set is complete.

Proof. The maximal solutions f satisfy $\|f(t)\| \leqslant A + |t|\, \|X_0\|$, where A is a constant. The details will not be given here. □

(b) Consider Newton's equation $d^2q/dt^2 = -\text{grad } U(q)$ of §7.1.5 again, and assume that the potential energy U is positive everywhere. Then every maximal solution $q(t)$ is defined on all of $\mathbf{R}$ (see V. Arnold).

Proof. Newton's equation may be written as $dq/dt = p$, $dp/dt = -\text{grad } U(q)$, where $q, p \in E$. First note that the total energy $H(q, p) = \frac{1}{2}\|p\|^2 + U(q)$ is a first integral by §7.4.1: $DH.X = \langle \text{grad } U(q), p\rangle - \langle p, \text{grad } U(q)\rangle = 0$.

Consider the maximal solution $(q(t), p(t))$ such that $q(0) = q_0, p(0) = p_0$, and put $H_0 = H(q_0, p_0)$. Since $\frac{1}{2}\|p\|^2 + U(q) = H_0$ and $U(q) \geqslant 0$, it follows that $\|dq/dt\| = \|p\| \leqslant \sqrt{(2H_0)}$, and so $\|q(t) - q_0\| \leqslant \sqrt{(2H_0)}\,|t|$. Thus if $|t| \leqslant T$, the solution $(q(t), p(t))$ remains in the compact set $\{(q, p) : \|q\| \leqslant \|q_0\| + \sqrt{(2H_0 T)},\ \|p\| \leqslant \sqrt{(2H_0)}\}$ and it is enough to apply Corollary 7.2.

Remarks

It can be shown that if there is a constant $k \geqslant 0$ such that $U(q) \geqslant -k\|q\|^2$, then every solution is defined on all of $\mathbf{R}$. However, if we take $E = \{q\} = \mathbf{R}$ and $U(q) = -q^4/2$, we see that the solution $q(t) = (t-1)^{-1}$ cannot be extended up to $t = 1$.

The reader will find other criteria for completeness in the book of R. Abraham and J. Marsden (p. 71).

7.5 ONE-PARAMETER GROUPS OF DIFFEOMORPHISMS

One-parameter groups of linear automorphisms of a Banach space and their connection with homogeneous linear constant-coefficient differential equations have been studied in §5.3. We shall now generalize that work.

7.5.1 One-parameter groups of diffeomorphisms

By a one-parameter group of diffeomorphisms $\phi_t (t \in \mathbf{R})$ of an open subset U of a Banach space E is meant a mapping ϕ: $\mathbf{R} \times U \to U$ such that:

(a) ϕ is of class C^r, $r \geqslant 1$.
(b) Given any $t \in \mathbf{R}$ the map $\phi_t : U \to U$ defined by $\phi_t(x) = \phi(t, x)$ is a diffeomorphism.
(c) The family ϕ_t, $t \in \mathbf{R}$, is a one-parameter group of transformations of

U: $\phi_0 = \mathrm{id}_U$; that is, $\phi_0(x) = x$ for all $x \in U$; $\phi_s \circ \phi_t = \phi_{s+t}$ for $s, t \in \mathbf{R}$. So:

$$\phi(s, \phi(t, x)) = \phi(t + s, x)$$

for all $s, t \in \mathbf{R}$ and all $x \in U$.

Note that the diffeomorphism inverse to ϕ_t is ϕ_{-t} for $\phi_t \circ \phi_{-t} = \phi_{t-t} = \phi_0 = \mathrm{id}_U$.

Let $x \in U$ be fixed. We may regard the curve $t \in \mathbf{R} \leadsto \phi_t(x) = \phi(t, x) \in U$ as defining the motion of a point whose position at the initial time 0 is $\phi(0, x) = x$ and at time t is $\phi_t(x)$. The orbit of this point is thus $\{\phi_t(x) : t \in \mathbf{R}\}$.

7.5.2 The generator of the group

By the velocity $X(x)$ of ϕ_t at a point x of U is meant the velocity at time $t = 0$ of $t \leadsto \phi_t(x) = \phi(t, x)$:

$$X(x) = \frac{\mathrm{d}}{\mathrm{d}t} \phi(t, x) \big|_{t=0}.$$

It follows that if ϕ is of class C^r, $r \geqslant 1$, we have defined a vector field X: $U \to E$ of class C^{r-1} whose integral curves are $t \leadsto \phi(t, x)$. We say that X is the (infinitesimal) generator of the one-parameter groups (which we also call a flow).

Example

The generator of the flow $\phi_t(x) = \mathrm{e}^{tA} x$ is $A \in \mathrm{End}(E)$.

Theorem 7.10

The velocity vector of the motion of a point at each instant is equal to the velocity vector of the flow at the position occupied by the point at the instant considered. In other words:

$$\frac{\mathrm{d}}{\mathrm{d}t} \phi_t(x) \big|_{t=s} = X[\phi_s(x)].$$

Proof.

$$\frac{\mathrm{d}}{\mathrm{d}t} \phi_t(x) \big|_{t=s} = \frac{\mathrm{d}}{\mathrm{d}t} \phi(s + r, x) \big|_{r=0} = \frac{\mathrm{d}}{\mathrm{d}t} \phi[r, \phi(s, x] \big|_{r=0} = X[\phi(s, x)]. \quad \square$$

The generator X of a flow is a complete vector field since the integral curves $t \leadsto \phi(t, x)$ are defined for all $t \in \mathbf{R}$. We shall show that, conversely, a complete vector field induces a flow. To sum up, as in §5.3, we shall have proved that there is a one-to-one correspondence between flows on U and complete vector fields on U.

Theorem 7.11

Let X be a complete vector field of class C^r, $r \geqslant 1$, on an open subset U of a Banach space E. For each $x \in U$, denote by $t \mapsto \phi(t, x)$ the maximal integral curve of X such that $\phi(0, x) = x$. Then $\phi: R \times U \to U$ is a one-parameter group of diffeomorphisms of U.

Proof. By Theorem 7.6, ϕ is of class C^r; condition (a) of §7.5.1 is thus satisfied.

We shall next verify condition (c). First, by definition, $\phi(0, x) = x$ for all x. Next, let $s, t \in \mathbf{R}$ and set $f(t) = \phi(t + s, x)$, $r = s + t$. Then:

$$\frac{\mathrm{d}}{\mathrm{d}t} f(t) = \frac{\mathrm{d}}{\mathrm{d}t} \phi(t+s, x) = \frac{\mathrm{d}}{\mathrm{d}t} \phi(r, x) = X[\phi(r, x)] = X[\phi(t+s, x)] = X[f(t)].$$

Thus f is the maximal integral curve of X such that $f(0) = \phi(s, x)$. The same is evidently true, by definition of ϕ, for $t \mapsto \phi(t, \phi(s, x))$. By uniqueness:

$$\phi(t+s, x) = \phi(t, \phi(s, x)).$$

As for condition (b) of §7.5.1, this is in fact a consequence of (a) and (c). Since ϕ is of class C^r, $r \geqslant 1$, the map $\phi_t : x \mapsto \phi(t, x)$ is of class C^r. We have seen that it is invertible and its inverse $(\phi_t)^{-1} = \phi_{-t}$ is of class C^r. It follows that ϕ_t is a C^r diffeomorphism of U on U.

Remark

If X is not complete, the last proof shows that for each $x_0 \in U$, there is a neighbourhood V of x_0 contained in U and an interval $]-a, a[$ such that:

(a) $\phi: \ ]-a, a[\ \times V \to U$ satisfies

$$\frac{\mathrm{d}}{\mathrm{d}t} \phi(t, x) = X[\phi(t, x)] \text{ for all } x \in V.$$

(b) For each $t \in \ ]-a, a[$, $\phi_t : x \mapsto \phi(t, x)$ is a diffeomorphism of V on $\phi(V)$.
(c) If $s, t, t+s \in \ ]-a, a[$ then $\phi_s \circ \phi_t = \phi_{s+t}$.

We summarize these properties by saying that ϕ defines a germ of a one-parameter group of diffeomorphisms in a neighbourhood of x_0.

Notes

The present chapter does not go deeply into the subject. The reader will find more complete accounts in the books by V. Arnold, E. A. Coddington and N. Levinson, and J. Rouche and J. Mawhin which have already been cited. Another proof of the existence theorem will be found in the book by H. Cartan. The books by R. Abraham and J.E. Marsden and by M.W. Hirsch and S. Smale contain recent advances in the global study of dynamic systems.

8

Conjugacy and local coordinates

This chapter deals with conjugacy in the group of diffeomorphisms. This concept is illustrated by, among others, the theorem of constant rank and the straightening theorem for vector fields.

PRELIMINARIES

In euclidean geometry we speak of figures which are congruent, that is, which can be superimposed by displacement. We prove that two figures are congruent by showing that they have in common a number of properties which are invariant under displacement (as in the case of congruent triangles). Similarly, two transformations of the space are regarded as equivalent if the action of one of them may be superimposed on the action of the other by a displacement. Thus two rotations of the euclidean plane are equivalent if their angles of rotation are equal.

We may extend this point of view to groups more general than the euclidean group.

By way of a second example, let us examine the group $GL(E)$ of automorphisms of a v.s. E. Two endomorphisms $A, B \in \mathrm{End}(E)$ are regarded as equivalent (we say they are 'similar') if there exists $S \in GL(E)$ such that $B = SAS^{-1}$.

How can we tell that A and B are similar? If we take $E = C^n$ and we confine ourselves to the dense open subset U of $\mathrm{End}(E)$ formed by the endomorphisms with distinct eigenvalues (see Appendix E), $A \in U$ and $B \in U$ are similar if, and only if, they have the same spectrum $\{\lambda_1, \ldots, \lambda_n\}$. For if S is the automorphism which sends the basis of E formed by the eigenvectors of A on to that formed by the eigenvectors of B, we have $B = SAS^{-1}$.

This classification provides a method of simplifying operations on $\mathrm{End}(E)$, provided that these operations commute with conjugation. Thus the endomorphism A above is similar to the endomorphism B whose matrix in the canonical basis of C^n is $\mathrm{diag}(\lambda_r)$. The question of raising to the kth power, k an integer, is then decidedly simplified:

$$A^k = S[\mathrm{diag}(\lambda_r)]^k S^{-1} = S\,\mathrm{diag}(\lambda_r^k) S^{-1}.$$

If the eigenvalues λ_r are positive, this formula gives, as a bonus, the possiblity of extending the operation of raising to the kth power to the case in which k is an arbitrary real number.

In this chapter we shall be interested in the group of C^k diffeomorphisms of a Banach space E. We shall introduce the corresponding notion of equivalence (C^k-conjugacy) and shall look for the invariants which will enable us to decide if two morphisms of E are equivalent.

8.1 C^k-CONJUGACY AND COORDINATES

8.1.1 C^k-conjugacy

Let f be a C^k map, $k \geqslant 1$, of a Banach space E in a Banach space E', f_1 a C^k map of a Banach space E_1 to a Banach space E_1'. We say that f and f_1 are C^k-conjugates if there are C^k diffeomorphisms $\phi : E \to E_1$ and $\phi' : E' \to E_1'$ such that the diagram in Figure 8.1(a) commutes; in other words, if $f_1 = \phi' \circ f \circ \phi^{-1}$.

8.1.2 Local C^k-conjugacy

The last property can only be verified locally. Let us retain the previous notation. Let $a \in E$, $a_1 \in E_1$; put $a' = f(a)$, $a_1' = f_1(a_1)$. We say that f and f_1 are locally C^k-conjugate in the neighbourhood of a and a_1 if there are open neighbourhoods of a, a_1, a', a_1', written as U, U_1, U', U_1' respectively, such that:

(a) $f(U) \subset U'$, $f_1(U_1) \subset U_1'$.

(b) The restrictions of f to U and of f_1 to U_1 are C^k-conjugate (Figure 8.1(b)).

Since the composition of diffeomorphisms is a diffeomorphism, local C^k-conjugacy is an equivalence relation.

Example

The diagram in Figure 8.1(c) shows that a C^k diffeomorphism f: $E \to E'$ is C^k-conjugate to the identity map.

We shall express all this in terms of coordinates.

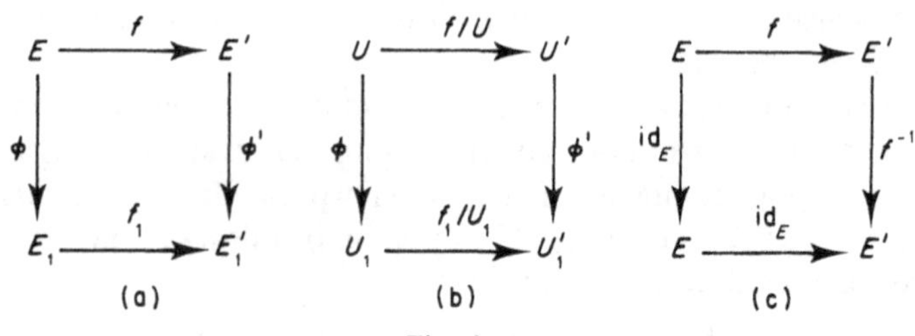

Fig. 8.1

8.1.3 Coordinates

Let U be an open subset of a finite-dimensional v.s. E on K (we shall often take $K = \mathbf{R}$, to fix the notation). If $\{e_k\}$ is a basis of E, each element u of U may be written in a unique way as $\sum u_k e_k$. Let x_k be the map $u \rightsquigarrow u_k$ of U into K. The functions $x_1, \ldots, x_n$ are the components of a map ϕ of U to K^n, which is the restriction to U of an isomorphism of E on K^n:

$$\phi : U \to K^n$$

$$u \to (u_1 = x_1(u), \ldots, u_n = x_n(u)).$$

We say that $(x_1, \ldots, x_n)$ is a linear coordinate system on U.

Of course $\phi : U \to \phi(U)$ is a diffeomorphism. This leads us to generalize the notion of coordinates.

We shall call a *chart with domain* U the pair (U, ϕ) formed by an open set U and a diffeomorphism ϕ of U on an open set of K^n. The components $(x_1, \ldots, x_n)$ of ϕ are called a *system of coordinates* and $(x_1(u), \ldots, x_n(u))$ are called the coordinates of $u \in U$ in the chart considered. The inverse map $\phi^{-1} : \phi(U) \subset K^n \to U$ is called a *parametrization* of U.

Example

Take $E = \mathbf{R}^2 = \{(x, y)\}$, $U = \{(x, y) : x > 0, y > 0\}$. Then $\phi : u = (x, y) \rightsquigarrow (\rho = [x^2 + y^2]^{1/2}, \theta = \tan^{-1}(y/x)) \in \mathbf{R}^2$ is a chart on U. We call $(\rho(u), \theta(u))$ the *polar coordinates* of u.

The inverse image under ϕ of the lines $\rho = \text{constant}$ in $\mathbf{R}^2$ are circles. This is why arbitrary coordinates are sometimes called *curvilinear coordinates.*

The image under ϕ of a figure in U is called the *reading* of this figure *in the chart.* The circles with centre 0 in $E = \mathbf{R}^2$ are thus read as lines in polar coordinates.

If f: $U \to K$ is a function, its expression in the new coordinates (x_k) (or, if we prefer, its reading in the chart (U, ϕ)) is, by definition, $f \circ \phi^{-1}$.

Now let f be a C^k map of an open subset U of a Banach space E of dimension n to an open subset U' of an m-dimensional Banach space E'. Let ϕ: $U \to K^n$ be a chart on U (with coordinates x_r), $\phi' : U' \to K^m$ a chart on U' (with coordinates y_s). The map $\phi' \circ f \circ \phi^{-1}$ is called the expression (or the reading) of f in the charts (U, ϕ) and (U', ϕ'). This is a C^k map of the open subset $\phi(U)$ of K^n to the open subset $\phi'(U')$ of K^m which, by construction, is C^k-conjugate to the map f. It thus provides a *model* of f in its C^k-conjugacy class.

8.1.4 Local diffeomorphisms and local coordinates

Recall that a C^k map f, $k \geqslant 1$, of an open subset U of a Banach space E to a Banach space F is called a local diffeomorphism at $a \in U$ if there exists an open neighbourhood V of a in U such that the restriction of f to V is a diffeomorphism of V on $f(V)$.

The inverse function theorem (Theorem 3.1) may thus be reformulated as follows: f is a local diffeomorphism at a if, and only if, the derivative $Df(a)$ is an isomorphism of E on F.

Let us now 'localize' the notions of §8.1.3.

Definition 8.1

Let E be an n-dimensional v.s. over K, U an open subset of E, and let a be a point of U. By a local coordinate system at a is meant the specification of n functions $x_i : U \to K$ of class C^k such that $u \in U \mapsto (x_1(u), \quad \ldots, x_n(u)) \in K^n$ is a local C^k-diffeomorphism at a.

From the inverse function theorem, in the form in which we have just given it, we immediately deduce the following result.

Theorem 8.1

$(x_1, \ldots, x_n)$ is a local coordinate system at a if, and only if the determinant of the matrix $(D_i x_j(a))$ is non-zero; or if, and only if, the linear forms $Dx_1(a), \ldots, Dx_n(a)$ form a basis of the dual space E^*.

The notion of local coordinates enables us to express that of local C^k-conjugacy in a convenient fashion. We retain the notation §8.1.2 and assume that E and E' are finite-dimensional. Then f and f_1 are locally C^k-conjugate in neighbourhoods of a and a_1 if their expressions in suitable charts are the same.

Example

Let E and F be v.s of the same dimension n, let U be an open subset of E, and let f: $U \to F$ be a local C^k-diffeomorphism at $a \in U$. Choose local coordinates $(y_1, \ldots, y^n)$ at $b = f(a)$. Then $(x_1 = y_1 \circ f, \ldots, x^n = y^n \circ f)$ is a local coordinate system at a, for the composition of two diffeomorphisms is a diffeomorphism. It follows that f, read in the coordinates (x_r), (y_s) is the identity map (compare with the example of §8.1.2).

We shall now generalize this result, assuming merely that $Df(a)$ is an isomorphism of E on to F.

8.2 LOCAL REPRESENTATION OF A DIFFERENTIABLE MAP

Unless otherwise stated, the Banach spaces E and F of this section are of finite dimensions n and m over the same field K.

8.2.1 The case in which $Df(a)$ is surjective

Theorem 8.2

Let U be an open subset of E, f a C^k map, $k \geqslant 1$, of U to F, such that Df is surjective at $a \in U$. Then f is locally C^k-conjugate in neighbourhoods of a and $f(a)$ to the canonical projection:

$$K^n = K^m \times K^{n-m} \to K^m$$

$$(x_1, \ldots, x_m, \ldots, x_n) \to (x_1, \ldots, x_m)$$

In other words, the local behaviour of f in the neighbourhood of a is that of its derivative $Df(a)$.

Proof. Since $Df(a) \in \mathscr{L}(E; F)$ is surjective, $n = \dim E \geqslant \dim F = m$. Let $(y_1, \ldots, y_m)$ be a local coordinate system at $b = f(a)$. Set $x_r = y_r \circ f$ for $r = 1, \ldots, m$. Then $Dx_r(a) = Dy_r(b) \circ Df(a)$. Since $Df(a)$ is surjective and $\{Dy_r(b)\}$ forms a basis of F^* (Theorem 8.1), the linear forms $Dx_1(a), \ldots, Dx_m(a)$ are linearly independent. Complete these by $n - m$ linear forms $x_{m+1}, \ldots, x_n$ so as to form a basis of E. Then, by Theorem 8.1, $(x_1, \ldots, x_n)$ is a local coordinate system at a.

By construction, the expression of f in these coordinates is:

$$(x_1(u), \ldots, x_n(u)) \mapsto (x_1(u), \ldots, x_m(u)). \qquad \square$$

Remark

The theorem holds even in infinite dimensions, provided that there is a closed subspace E_2 of E such that E is the direct sum $\ker Df(a) \oplus E_2$. This has been proved, under this hypothesis, in Theorem 3.3 in the form of an existence theorem rather than as a local model, as above.

8.2.2 The case in which $Df(a)$ is injective

Theorem 8.3

Let U be an open subset of E and f a C^k map of U to F, $k \geqslant 1$. Suppose that Df is injective at $a \in U$. Then f is locally C^k-conjugate in neighbourhoods of a and $f(a)$ to the canonical injection:

$$K_n \to K_m$$

$$(x_1, \ldots, x_n) \mapsto (x_1, \ldots, x_n, 0, \ldots, 0).$$

In other words, the local behaviour of f near a is again that of its derivative $Df(a)$.

Proof. Since $Df(a) \in \mathscr{L}(E; F)$ is injective, $n = \dim E \leqslant \dim F = m$. Let $(x_1, \ldots, x_m)$ be a local coordinate system at a. If S is a supplement of $Df(a)E$

in F, let $(x'_{n+1}, \ldots, x'_m)$ be a system of linear coordinates on S. Then $(x_1, \ldots, x_n, x'_{n+1}, \ldots, x'_m)$ is a local coordinate system on $U \times S$ in a neighbourhood of $(a, 0)$.

Define ϕ: $U \times S \to F$ by $\phi(u, s) = f(u) + s$. The derivative of ϕ at the point $(a, 0)$, evaluated at $(h, k) \in E \times S$, is $D\phi(a, 0)(h, k) = Df(a)h + k$; it is thus an isomorphism of $E \times S$ on F.

Define functions $y_1, \ldots, y_m$ in a neighbourhood of $b = f(a) \in F$ by $y_i = x_i \circ \phi^{-1}$ for $i = 1, \ldots, n$ and $y_i' = x_i' \circ \phi^{-1}$ for $i = n+1, \ldots, m$. Since $D\phi(a, 0)$ is an isomorphism and the linear forms $Dx_1(a), \ldots, Dx_n(a)$, $Dx'_{n+1}(a), \ldots, Dx'_m(a)$ are linearly independent, it follows that the $Dy_i(b)$ form a basis of F^*. By Theorem 8.1, $(y_1, \ldots, y_m)$ is therefore a local coordinate system on F in a neighbourhood of $b = f(a)$.

By construction, $y_i \circ f(u) = x_i(u)$ for $i = 1, \ldots, n$ and $y_i \circ f(u) = y_i \circ \phi(u, 0) = x_i'(0) = 0$ for $i = n+1, \ldots, m$. The expression of f in the coordinates (x_i), (y_i) is thus:

$$(x_1(u), \ldots, x_n(u)) \mapsto (x_1(u), \ldots, x_n(u), 0, \ldots, 0). \qquad \square$$

Theorem 8.3 may be extended to infinite-dimensional Banach spaces, provided that the subspace $Df(a)E$ of F is closed and has a closed supplement S.

The proof is the same: ϕ: $U \times S \to F = Df(a)E \oplus S$ defined by $\phi(u, s) = f(u) + (0, s)$ is a local diffeomorphism in a neighbourhood of $(a, 0)$. If g is the inverse local diffeomorphism, defined in a neighbourhood of $(f(a), 0)$, then $g \circ f$ maps an open neighbourhood V, of a in U, into $Df(a)E \times \{0\}$ and induces a diffeomorphism of the neighbourhood on an open neighbourhood of 0 in $Df(a)E$.

8.2.3 The rank of $Df(a)$

Recall that the rank of a linear map L: $E \to F$ is the dimension of the image of E.

We use the notation of §8.1.2. If f and f_1 are C^k-conjugate, $f_1 = \phi' \circ f \circ \phi^{-1}$. Thus $Df_1(a_1) = D\phi'(a') \circ Df(a) \circ D\phi^{-1}(a_1)$. As $D\phi'(a')$ and $D\phi^{-1}(a_1)$ are isomorphisms, the rank of Df at a is equal to that of Df_1 at the point $a_1 = f(a)$. In other words, the rank is invariant under C^k-conjugation.

It follows that f can be locally C^k-conjugate, in a neighbourhood of a, to a linear map f_1 (whose rank is plainly constant) only if the rank of Df is constant in a neighbourhood of a.

Thus f: $t \in \mathbf{R} \mapsto t^3 \in \mathbf{R}$ cannot be C^k-conjugate to a linear map in a neighbourhood of zero, for $f'(0) = 0$, while $f'(t) = 3t^2 \neq 0$ if $t \neq 0$.

It is thus necessary that the rank of Df should be constant in a neighbourhood of a in Theorems 8.2 and 8.3. Let us examine the reason for this.

Although the map $u \mapsto \operatorname{rank} Df(u)$ is clearly not continuous, it is lower semi-continuous: if $\operatorname{rank} Df(a) = p$, then $\operatorname{rank} Df(u) \geqslant p$ in a neighbourhood of a. For some minor of order p of the Jacobian matrix of f is non-zero at a, and since it depends continuously on u, it remains non-zero in a neighbourhood of a.

When the rank of $Df(a)$ is equal to the dimension of E (injectivity) or of (surjectivity), it takes its maximum value $\sup\{\dim E, \dim F\}$ at the point a. By semi-continuity, it thus remains constant in a neighbourhood of a. That is why the natural generalization of Theorems 8.2 and 8.3 is the following theorem.

8.2.4 The constant rank theorem

Theorem 8.4

In order that a C^k map $f\colon U \subset E \to F$, $k \geqslant 1$, be locally C^k-conjugate in a neighbourhood of $a \in U$ to a linear map, it is necessary and sufficient that the rank of $Df(u)$ be equal to a constant r in a neighbourhood of a.

Proof. Necessity is clear. Since a linear map is equal to its derivative, its rank is constant. And we have just seen (§8.2.2) that the rank is invariant under C^k-conjugation.

Let us prove sufficiency. We may assume that $a = 0$, $f(a) = 0$. On E and F choose linear coordinates (x_i') and (y_j') such that $Df(a)$ is given by $(x_1', \ldots, x_n') \rightsquigarrow (x_1', \ldots, x_r', 0, \ldots, 0)$ and identify E with $\mathbf{R}^n$ and F with $\mathbf{R}^m$ by means of these coordinates.

If $f_1, \ldots, f_m$ are the components of f, consider the map $\phi\colon U \to \mathbf{R}^n$ defined by $\phi = (f_1, \ldots, f_r, x_{r+1}', \ldots, x_n')$. The expression for $Df(0)$ shows that $D\phi(0)$ is the identity. By the inverse function theorem (Theorem 8.1), it follows that $x_1 = f_1, \ldots, x_r = f_r$, $x_{r+1} = x_{r+1}', \ldots, x_n = x_n'$ is a local coordinate system in a neighbourhood of $a = 0$. Let us write f in the coordinates (x_i), (x_j'); we find $(x_1, \ldots, x_n) \rightsquigarrow (x_1, \ldots, x_r, F_{r+1}(x), \ldots, F_m(x))$, where F_j is a C^k function of $x = (x_1, \ldots, x_r)$. Hence the Jacobian matrix of this contains the unit $r \times r$ matrix as its upper left-hand $r \times r$ block. Since its rank in a neighbourhood of $a = 0$ is equal to r, every $(r+1) \times (r+1)$ block has zero determinant. Thus $D_iF_j(x_1, \ldots, x_n) = 0$ for $i, j > r$. The functions F_j therefore do not depend on $x_{r+1}, \ldots, x_n$.

Set $y_1 = y_1', \ldots, y_r = y_r'$, $y_{r+1} = y_{r+1}' - F_{r+1}(y_1, \ldots, y_r), \ldots$, $y_m = y_m' - F_m(y_1, \ldots, y_r)$. We obtain a new local coordinate system (y_i) in a neighbourhood of $f(a) = 0$, for these formulae are invertible; i.e. $y_i' = y_i + F_i(y_1, \ldots, y_r)$ for $i > r$. Express f in the coordinates (x_i), (y_j); we find that $(x_1, \ldots, x_n) \rightsquigarrow (x_1, \ldots, x_r, 0, \ldots, 0)$. This is what we had to prove. □

As an exercise, the reader may prove the following general result which no longer assumes that the rank is constant.

Let f be a C^k map, $k \geqslant 1$, of an open subset of $\mathbf{R}^n$ to $\mathbf{R}^m$. Suppose that at some point a of this open set, the rank of $Df(a)$ equals r. Then f is locally

C^k-conjugate in a neighbourhood of a to:

$$\mathbf{R}^n = \mathbf{R}^r \times \mathbf{R}^{n-r} \to \mathbf{R}^m = \mathbf{R}^r \times \mathbf{R}^{m-r}$$
$$(u, v) \quad \to \quad (u, h(u, v))$$

where h: $\mathbf{R}^n \to \mathbf{R}^{m-r}$ is of class C^k and $Dh(0) = 0$. (See R. Abraham and J. Robbin, *Transversal Mappings and Flows*, Benjamin, New York, 1967, pp. 4 and 5).

8.3 THE MORSE-PALAIS LEMMA

Theorem 8.2 shows that a real-valued C^k function defined on an open subset of $\mathbf{R}^n$ is locally C^k-conjugate to an affine map in a neighbourhood of a point $a \in U$ where $Df(a) = 0$.

We shall investigate what happens near a point at which the derivative is zero.

8.3.1 Non-degenerate critical points

Let U be an open subset of a Banach space, E, and let f: $U \to \mathbf{R}$ be a function of class C^k, $k \geqslant 1$. We say that $a \in U$ is a *critical point* of f if $Df(a) = 0$.

Suppose that $k \geqslant 2$. The second derivative $D^2f(a)$, called the *hessian* of f at a, is a continuous symmetric bilinear form on E (see §4.1.3). But $\mathscr{L}(E, E; R)$ is canonically isomorphic to $\mathscr{L}(E; \mathscr{L}(E; R)) = \mathscr{L}(E; E^*)$ by Theorem A.8. Under this isomorphism $D^2f(a)$ is mapped to the continuous linear map $h \in E \rightsquigarrow D^2f(a)(h, .) \in E^*$. If this continuous linear map is an isomorphism of E on its dual E^*, we say that $D^2f(a)$ is *non-singular* and that the critical point a is *non-degenerate*.

Example

If $E = \mathbf{R}$, the point a is a critical point if $f'(a) = 0$. It is non-degenerate if $f''(a) \neq 0$.

Note that if ϕ is a C^k-diffeomorphism of a neighbourhood of a critical point a of f in E, then $D[f \circ \phi](\phi^{-1}(a)) = Df(a) \circ D\phi(\phi^{-1}(a)) = 0$. hence $\phi^{-1}(a)$ is a critical point of $f \circ \phi$. In other words, *C^k-conjugation preserves critical points.*

Suppose that a is a non-degenerate critical point of f. We shall examine the critical point $\phi^{-1}(a)$ of $f \circ \phi$. We may assume, for simplicity of notation, that $a = \phi(a) = 0$. If h, $l \in E$ then by Theorem 4.4:

$$D^2(f \circ \phi)(a)(h, l) = Df[\phi(a)]D^2\phi(a)(h, l) + D^2f[\phi(a)]\,(D\phi(a)h, D\phi(a)l)$$

Since a is a critical point, this reduces to:

$$D^2(f \circ \phi)(0)(h, l) = D^2f(0) \,.\, (D\phi(0)h, D\phi(0)l)$$

As $D\phi(0)$ is an isomorphism, this formula shows that $D^2(f\circ\phi)(0)$ is non-singular. The critical point $\phi^{-1}(a)=0$ is thus non-degenerate. Hence C^k-conjugation preserves non-degeneracy.

We shall now show that a C^3 function f: $\mathbf{R}^n \to R$ behaves, near a non-degenerate critical point a, like the part $f(a)+\frac{1}{2}D^2f(a)(h,h)$ of its Taylor series $f(a+h)=f(a)+\frac{1}{2}D^2f(a)(h,h)+o(\|h\|^2)$. In fact we shall prove this in the more general context of real Hilbert spaces, for the proof which follows, due to R. Palais, is decidedly simpler than that originally given by M. Morse and which applied only in $\mathbf{R}^n$.

We shall assume, without loss of generality, that the critical point a is the origin of E and that $f(a)=0$.

8.3.2 Statement of the Morse–Palais lemma

Let f be a real-valued function f of class C^{k+2}, $k \geqslant 1$, defined on a neighbourhood U of the origin in a Hilbert space E.

Suppose that $f(0)=0$ and that 0 is a non-degenerate critical point. Then there is a local C^k-diffeomorphism ϕ in a neighbourhood of 0, such that $\phi(0)=0$ and $f(x)=\frac{1}{2}D^2f(0)(\phi(x),\phi(x))$.

We begin with a lemma.

Lemma 8.1

Let E be a Banach space. If $B \in \mathrm{End}(E)$ is such that $\|1-B\|<1$, then there is a sequence of polynomials in B which converges in the Banach space $\mathrm{End}(E)$ to an automorphism R such that $R^2=B$, and which we shall denote by $\sqrt{B}$.

Moreover, there is a neighbourhood U of $1=\mathrm{d}_E$ such that $B\in U \rightsquigarrow \sqrt{B}$ is a C^∞ diffeomorphism of U on its image.

Proof. A possible square root R of B must satisfy $S=\frac{1}{2}[(1-B)+S^2]$, if we put $R=1-S$. To prove its existence, we shall use a method of successive approximation. Define a sequence $A_n \in \mathrm{End}(E)$ recursively:

$$A_0=0, A_{n+1}=[(1-B)+A_n^2]/2, n=0,1,\ldots.$$

Set $a=\|1-B\|$.

(a) We first prove that $\|A_n\|<\sqrt{a}$. This is clear when $n=0$. Assume it has been proved for n; it follows for $n+1$ since:

$$\|A_{n+1}\| \leqslant [\|1-B\|+\|A_n\|^2]/2 \leqslant (a+a)/2=a$$

which is less than $\sqrt{a}$ since $0<a<1$.

(b) A_n is a polynomial in $(1-B)$, and thus in B. This is clear, by induction on n, starting from the definition of A_n.

(c) The sequence A_n converges in $\mathrm{End}(E)$. We have $A_{n+1}-A_n=(A_n^2-A_{n-1}^2)/2$ and similar relations with n replaced by $n+1,\ldots,n+p-1$. Add these term by term: $A_{n+p}-A_n=(A_{n+p-1}^2-A_{n-1}^2)/2$. By (b), the A_i

commute. We may thus write:

$$A_{n+p} - A_n = (A_{n+p-1} - A_{n-1})(A_{n+p-1} + A_{n-1})/2.$$

Since $\| A_i \| < \sqrt{a}$, it follows that $\| A_{n+p} - A_n \| < (\sqrt{a}) \| A_{n+p-1} - A_{n-1} \|$, which shows that $\| A_{n+p} - A_n \| \leqslant (\sqrt{a})^n \| A_p - A_0 \| < (\sqrt{a})^{n+1}$. As $0 < \sqrt{a} < 1$, we see that A_n is a Cauchy sequence, which thus converges to an element S of the complete space End(E).

(d) $(1 - S)^2 = B$. It is enough to let $n \to +\infty$ in the relation which defines A_n recursively. Note that, since $\| A_n \| < \sqrt{a}$, we have $\| S \| < \sqrt{a}$. In other words, $R = 1 - S$ satisfies $\| 1 - R \| < \sqrt{a}$. Then $R = \sqrt{B}$ lies in the ball $B(1, \sqrt{r})$ if B is in the ball $B(1, r)$ of centre 1 and radius $r < 1$ in End(E).

(e) The derivative of the map Q: End(E) $\to$ End(E) of class C^∞, defined by $Q(A) = A^2$, is $DQ(A)h = Ah + hA$ for $h \in$ End(E). It follows that $DQ(1)$)is invertible. By the inverse function theorem and Theorem 4.6, Q is thus a C^∞ diffeomorphism of a neighbourhood U of $1 = \mathrm{id}_E$ on to its image. Now take $r < 1$ so small that $B(1, \sqrt{r}) \subseteq U$. Then by (d,)the restriction of Q^{-1} to $B(1, r)$ is simply the map $B \mapsto R = \sqrt{B}$ that we have constructed. This shows the uniqueness of the square root and proves that $B \mapsto \sqrt{B}$ is C^∞. □

8.3.3 Proof of the Morse–Palais lemma

We may assume that U is an open ball with centre 0. This enables us to apply the fundamental theorem of integral calculus, first to f, then to Df:

$$f(x) = f(x) - f(0) = \int_0^1 Df(tx)x \,\mathrm{d}(t),$$

$$Df(tx) = Df(tx) - Df(0) = \int_0^1 D^2 f(stx)tx \,\mathrm{d}s.$$

Hence

$$f(x) = \int_0^1 \int_0^1 D^2 f(stx)(x, x)t \,\mathrm{d}s \,\mathrm{d}t.$$

By §2.6.4(d) we may thus write $f(x) = (g)(x)(x, x)$, where

$$g(x) = \int_0^1 \int_0^1 D^2 f(stx)t \,\mathrm{d}s \,\mathrm{d}t. \tag{8.1}$$

Hence g is a C^k map of U to the Banach space $\mathscr{L}(E, E; \mathbf{R})$, which we know is canonically isomorphic to $\mathscr{L}(E; \mathscr{L}(E; \mathbf{R})) = \mathscr{L}(E; E^*)$ (Appendix A). Moreover, the scalar product $\langle , \rangle$ of E induces a canonical isomorphism $h \in E \mapsto \langle h, . \rangle \in E^*$. Thus $\mathscr{L}(E, E; \mathbf{R})$ is canonically isomorphic to $\mathscr{L}(E; E) =$ End(E). The image $A(x)$ of $g(x)$ under this isomorphism is given by:

$$g(x)(h, l) = \langle A(x)h, l \rangle \quad \text{for} \quad h, l \in E. \tag{8.2}$$

In particular:

$$\tfrac{1}{2}D^2f(0)(x, x) = g(0)(x, x) = \langle A(0)x, x\rangle \tag{8.3}$$

which shows that $A(0)$ is an isomorphism, since 0 is a non-degenerate critical point. Thus:

$$f(x) = \langle A(x)x, x\rangle \quad \text{for} \quad x \in U. \tag{8.4}$$

We are looking for a local diffeomorphism ϕ such that $\phi(0) = 0$ and $f(x) = \tfrac{1}{2}Df(0)(\phi(x), \phi(x))$; that is, by equations (8.3) and (8.4):

$$\langle A(x)x, x\rangle = \langle A(0)\phi(x), \phi(x)\rangle. \tag{8.5}$$

To exhibit it, set $B(x) = A(0)^{-1}A(x)$. Since g is continuous, so is A, by equation (8.2). Making U smaller if necessary, we may therefore assume that $A(0)^{-1}A(x)$ is close to $A(0)^{-1}A(0) = 1$; that is, $\| 1 - B(x) \| \leqslant 1$ for $x \in U$.

By the last lemma, $B(x)$ has a square root $R(x)$. We shall show that $\phi(x) = R(x)x$ has all the required properties.

(a) *ϕ is of class C^k*. For as g is of class C^k, so are A and $B = A(0)^{-1}A$, by equation (8.2). As the 'square root' map is C^∞, the composite map $x \rightsquigarrow B(x) \rightsquigarrow R(x) = \sqrt{B(x)}$ is of class C^k.

(b *ϕ is a local diffeomorphism and $\phi(0) = 0$*. By Leibniz' rule, $D\phi(x)h = DR(x)(h, x) + R(x)h$ for $h \in E$. Thus $D\phi(0) = R(0) = \sqrt{B(0)} = 1$ is an isomorphism. It is therefore enough to apply the inverse function theorem. Also $\phi(0) = R(0)0 = 0$.

(c) *ϕ satisfies equation* (8.5). By Schwarz' theorem (Theorem 4.1), $D^2f(\,)$ is symmetric. By equation (8.1), so is $g(x)$. It follows from equation (8.2) that $A(x)$ is equal to its adjoint $A(x)^*$. hence $B(x)^* = [A(0)^{-1}A(x)]^* = A(x)A(0)^{-1}$; and so $B(x)^*A(0) = A(x) = A(0)B(x)$.

In the last relation we may replace $B(x)$ by an arbitrary polynomial in $B(x)$. As $R(x) = \sqrt{B(x)}$ is the limit of a sequence of such polynomials (by the last lemma), we also have $R(x)^*A(0) = A(0)R(x)$. Finally:

$$\begin{aligned}\langle A(0)\phi(x), \phi(x)\rangle &= \langle A(0)R(x)x, R(x)x\rangle \\ &= \langle R(x)^*A(0)R(x)x, x\rangle = \langle A(0)R^2(x)x, x\rangle \\ &= \langle A(0)B(x)x, x\rangle = \langle A(x)x, x\rangle.\end{aligned}$$

□

Remark

The reader will find in R. Abraham and J. Marsden (pp. 175–176) an even shorter proof which, however, requires ideas not covered in this book.

Corollary 8.1

The non-degenerate critical points of f are isolated.

Corollary 8.2

Let f: $\mathbf{R}^n \to \mathbf{R}$ be a function of class C^{k+2}, $k \geqslant 1$. Suppose that $f(0) = 0$ and that 0 is a non-degenerate critical point of f. Then on some neighbourhood of 0 there are local coordinates (x_i) in which f may be expressed as $x_1^2 + \ldots + x_r^2 - x_{r+1}^2 - \ldots - x_n^2$.

In particular, the local C^k-conjugacy class at 0 of f is characterized by the number of squares with a + sign which occur in the decomposition of the quadratic form $D^2f(0)(h, h)$ as a sum of squares.

The reader will find other applications of the Morse–Palais lemma in Chapter 10.

8.4 LINEARIZATION OF VECTOR FIELDS

8.4.1 Conjugacy in the group of diffeomorphisms

We return to the C^k-conjugacy defined in §8.1.1 and to the notation used already: f: $E \to E'$ is C^k-conjugate to $f_1 : E_1 \to E_1'$ if there are C^k diffeomorphisms ϕ: $E \to E_1$ and $\phi' : E' \to E_1'$ such that $f_1 = \phi' \circ f \circ \phi^{-1}$.

This implies that all the C^k diffeomorphisms f: $E \to E$ are C^k-conjugates of the identity map (Figure 8.2). Thus C^k-conjugacy is a very coarse classification when E and E' coincide. We shall now refine it.

We say that two C^k maps f: $E \to E$ and $f_1 : E \to E$ are conjugate in the group $\mathrm{Diff}^k(E)$ of C^k-diffeomorphisms of E if there exists $\phi \in \mathrm{Diff}^k(E)$ such that $f_1 = \phi \circ f \circ \phi^{-1}$ (we usually simply say 'conjugate' if no ambiguity is possible).

Some new invariants appear. If a is a fixed point of f, then $\phi(a)$ is a fixed point of f_1. More generally, the image under ϕ of each orbit $\{f^n(a) : n \in N\}$ of f is an orbit of f_1. The number of orbits of period P (that is, containing exactly P distinct elements) is thus an invariant.

The reader will easily be able to define the corresponding local notion by comparison with §8.1.2.

8.4.2 Conjugate of a one-parameter subgroup of diffeomorphisms

Let F: $R \times E \to E$ be a one-parameter subgroup of diffeomorphisms of E, with

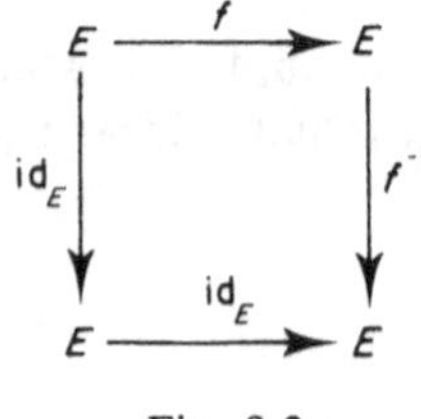

Fig. 8.2

parameter $t \in \mathbf{R}$ (see §7.5). It is easy to see that its transform $\phi \circ F_t \circ \phi^{-1}$ by $\phi \in \mathrm{Diff}^k(E)$ is also such a subgroup.

We now try to express its generator, written $\phi_* X$, in terms of the generator X of F. The chain rule shows that:

$$\frac{\mathrm{d}}{\mathrm{d}t}[\phi \circ F_t \circ \phi^{-1}(x)]\bigg|_{t=0} = D\phi(\phi^{-1}(x))\frac{\mathrm{d}}{\mathrm{d}t}F(t, \phi^{-1}(x))\bigg|_{t=0}.$$

Thus $(\phi_* X)(x) = D\phi(\phi^{-1}(x))X(\phi^{-1}(x))$. The vector field $\phi_* X$ is called the *image of X by ϕ*.

Note that since $D\phi$ is of class C^{k-1}, so is $\phi^* X$. This is an example of a C^{k-1} vector field which, however, induces a C^k flow $\phi \circ F_t \circ \phi^{-1}$. This does not, of course, contradict Theorem 7.6.

Given a one-parameter subgroup of diffeomorphisms F_t (or its generator X), we would like to find out whether there is a diffeomorphism ϕ such that $\phi \circ f_t \circ \phi^{-1}$ (or $\phi_* X$) has a simple form. In particular, can we choose ϕ so that $\phi \circ F_t \circ \phi^{-1}$ is a one-parameter subgroup of translations of E? This is certainly impossible in general: a translation has no fixed point, while F_t may well have one.

If a is a fixed point of F_t—that is, if $F(t, a) = a$ for all t—then $X(a) = 0$ (and conversely). We say that a is a *critical point* of X. A point of E at which X is non-zero is called a *regular point* of X; this agrees with the terminology introduced in §8.3.1.

The existence of a critical point of X is, as we have just seen, an obstacle to the transformation of F_t into a group of translations. We shall see that *locally* it is the only obstacle.

8.4.3 The straightening theorem

Theorem 8.5

Let a be a regular point of a C^k vector field X, $k \geqslant 1$, on an open subset Ω of a Banach space E. Then there is an open neighbourhood U of a and a C^k-diffeomorphism ϕ: $U \to \phi(U)$ such that $\phi_* X$ is constant on $\phi(U)$. In particular, if $E = \mathbf{R}^n$, there is a local coordinate system $(x_1, \ldots, x_n)$ at a in which the components of the field X are $X_1 = 1$, $X_2 = \ldots = X_n = 0$. In the coordinate system, $\phi \circ F_t \circ \phi^{-1}$ may be written as $(x_1, \ldots, x_n) \mapsto (x_1 + t, x_2, \ldots, x_n)$; this is a (local) group of translations.

Proof. Without loss of generality, we may assume that a is the origin 0 of E. Let E' be a closed subspace complementary to the space induced by $X(0)$: $E = \mathbf{R}X(0) \oplus E'$. Then $\Omega \cap E'$ is an open neighbourhood of 0 in E'.

Let $F(t, x)$ be the local flow defined by X (see the Remarks at the end of §7.5). Since F is continuous, we may choose an open neighbourhood U of 0 in $\Omega \cap E'$ and an interval $]-2r, 2r[$ such that $F(t, u) \in \Omega$ if $-2r < t < 2r$, $u \in U$. Set $I =]-r, r[$ and let G be the restriction of F to $I \times U$. Since X is of class C^k, so are F and G by Theorem 7.6. We find the derivative of G at $(0, 0)$. If

$(h,k)\in\mathbf{R}\times E'$ we have $DG(0,0)(h,k)=D_1G(0,0)h+D_2G(0,0)k$. But:

$$D_1G(s,u)=\frac{\mathrm{d}}{\mathrm{d}t}F(t,u)\,|_{t=s}=X[F(s,u)];$$

thus $D_1G(0,0)h=hX(0)$. Also $F(0,u)=u$ implies that $D_2G(0,0)k=k$. Hence $DG(0,0)$: $(h,k)\rightsquigarrow hX(0)+k$ is an isomorphism of $\mathbf{R}\oplus E'$ on E. By the inverse function theorem, and making r and U smaller if necessary, G is thus a C^k-diffeomorphism of $I\times U$ on $G(I\times U)$.

Set $\phi=G^{-1}$ and examine $\phi\circ F_t\circ\phi^{-1}=G^{-1}.\ F_t\circ G$ for $-r<t<r$ (chosen so as to remain in Ω). Since $F_t[F_s(u)]=F_{t+s}(u)$, we have:

$$\phi\circ f_t\circ\phi^{-1}:(s,u)\in I\times U\rightsquigarrow(s+t,u)\in\,]-2r,2r[\,\times U.$$

The image of F_t by ϕ is thus a translation $(s,u)\rightsquigarrow(s+t,u)$ of $\mathbf{R}\oplus E'\sim E$. Its generator $\phi{*}X$ is the constant field $\phi{*}X(0)=(1,0)$.

Remarks

(a) This theorem is called the straightening, or linearization, theorem because the orbits of $\phi\circ F_t\circ\phi^{-1}$ are parallel lines (Figure 8.3).

(b) The theorem about C^k dependence of $F_t(x)$ on x may be 'read' immediately so as to apply to the expression $\phi\circ F_t\circ\phi^{-1}$ of $F_t(x)$ in the chart $(I\times U,\phi)$.

(c) Another very ingenious proof is given by E. Nelson (see the Bibliography).

Corollary 8.2

A differentiable vector field X on $\mathbf{R}^n$ has $n-1$ first integrals, whose derivatives are linearly independent at each point of some neighbourhood of a regular point of X.

Proof. Let 0 be this regular point. In some neighbourhood of 0 there are local coordinates $(x_1,\ldots,x_n)$ in which the field $\phi{*}X$ has components $(1,0,\ldots,0)$. It follows that the coordinate functions $x_2,\ldots,x_n$ are first integrals of $\phi{*}X$

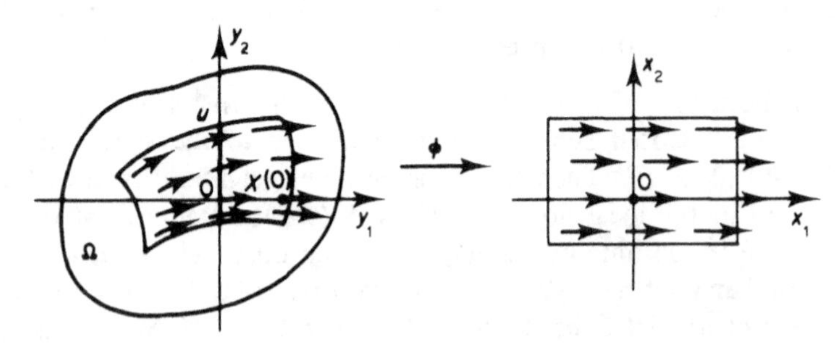

Fig. 8.3

(see §7.4.1). Clearly the derivatives of $x_2, \ldots, x_n$ are linearly independent at each point. Thus $x_i \circ \phi$, $i = 2, \ldots, n$, are $n-1$ first integrals of X, whose derivatives are linearly independent at each point. For by §§7.4.1 and 8.4.2, if $x = \phi(y)$ we have:

$$D(x_i \circ \phi)(y)X(y) = Dx_i(\phi(y))D\phi(y)X(y)$$
$$= Dx_i(x)D\phi(\phi^{-1}(x))X(\phi^{-1}(x)) = Dx_i(x)(\phi_* X(x) = 0. \quad \square$$

Remarks

It should not be thought that a vector field, defined on an arbitrary open subset of $\mathbf{R}^n$ in which it has no critical points, has $n-1$ first integrals defined on all of U and whose derivatives at each point are linearly independent.

Here is a counter-example if $n = 4$. In $\mathbf{R}^4 = \{m = (q_1, q_2, p_1, p_2)\}$ consider the vector field X with components $(p_1, p_2, -q_1, -2q_2)$. This has no critical points on the open set $U = \mathbf{R}^4 \setminus \{0\}$. It is easy to see (§5.6.1) that $f(m) = p_1^2 + q_1^2$ and $g(m) = p_2^2 + 2q_2^2$ are first integrals whose derivatives $Df(m)$ and $Dg(m)$ are linearly independent. However, there is no third (here $n-1=3$) first integral h such that $Df(m)$, $Dg(m)$ and $Dh(m)$ are linearly independent at all points of U. This is because the integral curve of X, issuing from an arbitrary point a of U, is everywhere dense in the subset $f^{-1}(a) \cap g^{-1}(a)$.

The reader may prove this by making use of the Jacobi–Kronecker theorem: each orbit of the rotation $z \rightsquigarrow e^{2\pi i r}z$ of the circle $\{z \in \mathbf{C} : |z| = 1\}$ is everywhere dense on the circle if $r \in \mathbf{R}$ is irrational (in our case $r = \sqrt{2}$).

With the aid of these ideas, the reader may construct a counter-example on $\mathbf{R}^3$: think of a toroidal bobbin on which is twisted a thread which describes an orbit.

It can be shown (W. Kaplan, 1940) that the integral curves of a differentiable vector field on $\mathbf{R}^2$ without critical points are the level curves $f^{-1}(r)$ of a function $f: \mathbf{R}^2 \to \mathbf{R}$. This function is thus a *global* first integral.

Notes

The problem of classification by conjugacy in groups of diffeomorphisms has motivated much contemporary research. The reader will find in the book by V. Arnold the classification of critical points of linear vector fields: $X: x \rightsquigarrow A(x)$, where $A \in \mathrm{End}(E)$; and in the book by E. Nelson that of critical points of arbitrary vector fields. Lastly, R. Abraham and J. Marsden, and M. Hirsch and S. Smale, provide much information on this topic.

9

Differentiable submanifolds

This chapter introduces the concepts of differentiable submanifolds, tangent spaces at points of a submanifold and differentiable maps from one submanifold to another.

9.1 DIFFERENTIABLE SUBMANIFOLDS

In the spirit of the last chapter, let us say that two subsets of a Banach space E are equivalent if one is obtained from the other by means of a diffeomorphism $\phi: E \to E$. From this point of view, the simplest subsets are those which are equivalent to a vector subspace or to an open subset of such a subspace. A piece of a curve or a surface in ordinary space is usually of this type. However, a circle or sphere are only locally homeomorphic to a line or a plane. If we wish to include these in our study, we must 'localize' the concept of equivalence under diffeomorphisms.

9.1.1. First definition of submanifolds

Let F be a closed subspace of a Banach space E. A subset M of E is called a *C^k submanifold of E modelled on F* if each point $m \in M$ has an open neighbourhood U such that there is a C^k diffeomorphism ϕ of U on $\phi(U)$ such that $\phi(U \cap M) = \phi(U) \cap F$.

Equivalently, (see §8.1) we may say that near each point $m \in M$ there is a chart (U, ϕ) such that $\phi(U \cap M) = \phi(U) \cap F$.

Having given this general definition, we shall confine ourselves in what follows to the case in which E is a real n-dimensional vector space. If the subspace F is of dimension d, we shall say that M is a submanifold of *E of dimension d* (or of *codimension $n - d$*).

Usually we shall omit the qualification 'C^k'; it is understood that $k \geqslant 1$.

Examples

(a) The unit circle S^1 is a C_∞ submanifold of $\mathbf{R}^2 = \{(x, y)\}$ of dimension 1 (see Figure 9.1). Here $\phi: (x, y) \rightsquigarrow (\tan^{-1} y/x,\ [x^2 + y^2]^{1/2} - 1)$.

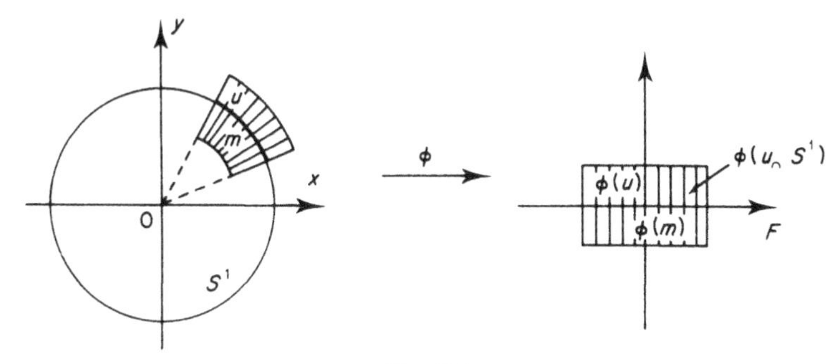

Fig. 9.1

(b) Every affine subspace of dimension d is a submanifold of dimension d.

Consequences

(a) If $d=n$, then $F=E$ and $\phi(U \cap M)=\phi(U) \cap E=\phi(U)$. Thus $U \cap M=U$ and $U \subset M$. The submanifolds of maximal dimension are hence open subsets of E (and conversely).

(b) If $d=0$, then $F=\{0\}$ and $\phi(U \cap M)=\phi(U) \cap F=\{0\}$. Hence $U \cap M$ reduces to the point $\phi^{-1}(0)$. The submanifolds of zero dimension are thus discrete subsets of E (and conversely).

(c) The image of a submanifold under a diffeomorphism f: $E \rightarrow E$ is a submanifold of the same dimension.

(d) Since ϕ induces a homeomorphism of $U \cap M$ on the open subset $\phi(U) \cap F$ of F, a submanifold of dimension d has, locally, the same topological properties as $\mathbf{R}^d$.

Examples

(a) Choose a point m in a submanifold M of dimension 1. Then a neighbourhood of $M \backslash \{m\}$ consists of two disjoint connected sets. This is why the subset of $\mathbf{R}^2=\{(x, y)\}$ with equation $xy=0$ is not a submanifold: with the origin removed, it has four connected components.

(b) The same reasoning applies to the image A of the curve f: $\mathbf{R} \rightarrow \mathbf{R}^2$ whose graph is given in Figure 9.2, and in which the arrow indicates that $f(t)$ tends to $f(0)$ as $t \rightarrow +\infty$. Near $f(0)$, $A \backslash \{f(0)\}$ has three connected components.

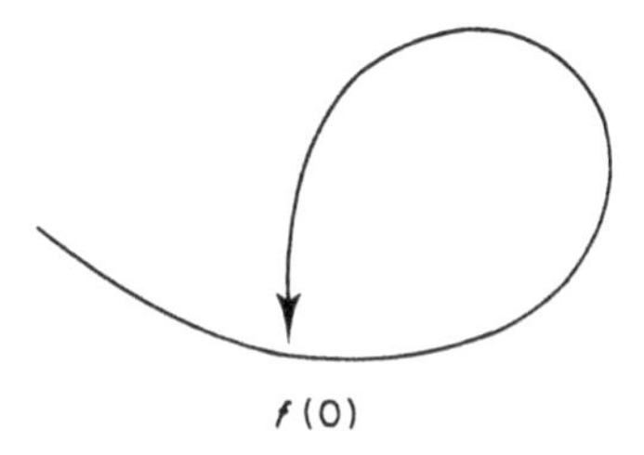

Fig. 9.2

9.1.2 Second definition of submanifolds

Theorem 9.1

A subset M of an n-dimensional vector space E is a d-dimensional submanifold if, and only if, every point $m \in M$ has an open neighbourhood U on which are defined $n-d$ functions $f_i : U \to \mathbf{R}$ of class C^k such that:
(a) $U \cap M$ is the set of common zeros of the functions.
(b) For all $u \in U$ the linear forms $Df_i(u)$, $i = 1, \ldots, n-d$, are linearly independent.

Proof. Necessity—Identify E with $\mathbf{R}^n = \mathbf{R}^d \times \mathbf{R}^{n-d}$ and F with $\mathbf{R}^d \times \{0\}$. If (U, ϕ) is the chart of the first definition given in §9.1.1, and if $(x_1, \ldots, x_n)$ are the canonical coordinates of $E = \mathbf{R}^n$, the functions $f_i = x_i \circ \phi$, $i = d+1, \ldots, n$, obviously satisfy (a) and (b) in Theorem 9.1.

Sufficiency—Let $m \in M$. Complete the linear forms $Df_i(m)$, $i = d+1, \ldots, n$, which are linearly independent, by adding the linear forms $f_1, \ldots, f_d$ so as form a basis of E^*. By Theorem 8.1, $(f_1, \ldots, f_n)$ is a local coordinate system on a neighbourhood $V \subset U$ of m. The map ϕ: $V \to \mathbf{R}^n$ with components $f_1, \ldots, f_n$ is a diffeomorphism of V on $\phi(V)$ such that $\phi(V \cap M) = \phi(V) \cap (\mathbf{R}^d \times \{0\})$.

Warning

Suppose that a subset M of E is the set of common zeros of $n-d$ functions g_i. It may happen that the derivatives of the g_i are not linearly independent on M but that nevertheless M is a submanifold. This simply means that the g_i are badly chosen.

Examples

(a) Let M be the set of common zeros of functions $g_1 : (x, y, z) \in \mathbf{R}^3 \mapsto x^2$ and $g_2 : (x, y, z) \in \mathbf{R}^3 \mapsto y^2$. The derivatives Dg_1 and Dg_2 vanish on M. However, M is a submanifold. We should have to choose $f_1(x, y, z) = x$ and $f_2(x, y, z) = y$.

(b) We have seen that the set A of zeros of $g_1 : (x, y) \in \mathbf{R}^2 \mapsto xy$ is not a submanifold. However, the fact that Dg_1 vanishes on A is not sufficient to prove this. There is no reason, *a priori*, to suppose that another function f_1 would not have the required properties.

We recall (see §2.5) that if g is a differentiable map of an open subset U of the Banach space E to a Banach space E', then a point $a \in U$ is said to be a *critical* point (resp. regular) of g if $Dg(a)$ is not (resp. is) surjective. A point $b \in E'$ is called a ***regular value*** of g if $g^{-1}(b)$ contains no critical point.

Corollary 9.1

Let g be a differentiable map of an open subset U of E to E'. Suppose that

$b \in E'$ is a regular value and $g^{-1}(b)$ is non-empty. Then $g^{-1}(b)$ is a submanifold of E of codimension dim E'.

Proof. Identify E' with $\mathbf{R}^p$ by a choice of linear coordinates. Let $(g_1, \ldots, g_p)$ be the components of g and $(b_1, \ldots, b_p)$ those of b. Set $f_i = g_i - b_i$. Then $g^{-1}(b) = \{x \in U : f_i(x) = 0,\ i = 1, \ldots, p\}$ and, since $Dg(x)$ is surjective, the $Df_i(x) = Dg_i(x)$ are linearly independent for all $x \in U$. It is now enough to use Theorem 9.1. □

9.1.3 Third definition of submanifolds

We have, in fact, shown that M is a d-dimensional submanifold in $E = \mathbf{R}^n$ if, and only if, each point $m \in M$ has an open neighbourhood U such that there is a map $g: U \to F = \mathbf{R}^{n-d}$ satisfying:
(a) $Dg(u)$ is surjective for $u \in U$.
(b) $M \cap U = g^{-1}(b)$ for $b \in F$.

Sufficiency has just been proved. Necessity is contained in Theorem 9.1.

9.1.4 Spheres

The sphere S^n in $\mathbf{R}^{n+1} = \{(x_1, \ldots, x_{n+1})\}$ is a submanifold of dimension n, since $S^n = f^{-1}(1)$, where $f: (x_1, \ldots, x_{n+1}) \in \mathbf{R}^{n+1} \mapsto x_1^2 + \ldots + x_{n+1}^2 \in \mathbf{R}$. Also $+1$ is a regular value of f, for $Df(x) = (2x_1, \ldots, 2x_{n+1})$ is never zero on S^n.

9.1.5 The unimodular group

Identify $A \in \mathrm{End}(\mathbf{R}^n)$ with its $n \times n$ matrix in the canonical basis. Then identify this matrix with an element of $\mathbf{R}^{n^2}$, reading its elements from left to right and from top to bottom (the Arabic reader will make the necessary adjustment).

The group GL(n) of automorphisms of $\mathbf{R}^n$ is an open subset of $\mathbf{R}^{n^2}$ for it is the inverse image of the open set $\mathbf{R}\backslash\{0\}$ under the continuous map $f: A \in \mathrm{End}(\mathbf{R}^n) \mapsto \det(A)$.

We shall prove that the unimodular group $\mathrm{SL}(n) = f^{-1}(1)$ consisting of those endomorphisms with determinant $+1$ is a submanifold of $\mathbf{R}^{n^2}$. The map f is C^∞, for $f(A)$ is a polynomial in the coefficients of A. As it is a homogeneous polynomial of degree n, $f(tA) = t^n f(A)$ for $t \in \mathbf{R}$. Hence $Df(tA)A = (\mathrm{d}/\mathrm{d}t)f(tA) = nt^{n-1}f(A)$. Put $t = 1$, so that $Df(A)A = nf(A)$. Thus $Df(A) \neq 0$ on $\mathrm{SL}(n) = f^{-1}(1)$. It follows that $+1$ is a regular value of f: $U \to \mathbf{R}$ and it is enough to apply Corollary 9.1.

9.1.6 The orthogonal group

A symmetric $n \times n$ matrix has at most $n(n+1)/2$ distinct elements. This enables us to identify the set S of symmetric endomorphisms s (that is, equal to their transpose ${}^ts$) with $\mathbf{R}^{n(n+1)/2}$, just as we have identified End($\mathbf{R}^n$) with $\mathbf{R}^{n^2}$.

Consider the map f of the open subset $U = \mathrm{GL}(n)$ of $\mathbf{R}^{n^2}$ to $S = \mathbf{R}^{n(n+1)/2}$ defined by $f(A) = {}^tAA$. This is clearly a C^∞ map. By Liebniz' rule, $Df(A)h = {}^thA + {}^tAh$ for $h \in \mathrm{End}(E)$. We claim that $Df(A)$ is surjective. For if $s \in S$, then $Df(A)h = s$ with $h = {}^t(A^{-1})s/2$. It follows that $\mathrm{id}_{\mathbf{R}^n} \in S$ is a regular value of f. As the orthogonal group $\mathrm{O}(n)$ of $\mathbf{R}^n$ is simply $f^{-1}(\mathrm{id}_{\mathbf{R}^n})$, Corollary 9.1 shows that $\mathrm{O}(n)$ is a submanifold of $\mathbf{R}^{n^2}$ of dimension $n^2 - n(n+1)/2 = n(n-1)/2$.

The proof of the following generalization of Corollary 9.1 is left to the reader.

Theorem 9.2

Let f be a C^k map of an open subset U of E to F. Suppose that $Df(u)$ is of constant rank r for $u \in U$. Then the inverse image $f^{-1}(b)$ (assumed to be non-empty) of a point $b \in F$ is a C^k submanifold of E, of dimension $\dim(E) - r$.

□

9.1.7 The graph of a map

Theorem 9.3

Let E_1 and E_2 be two finite-dimensional vector spaces, U_1 an open subset of E_1, and $g\colon U_1 \to E_2$ a C^k map. Then the graph $\{(u_1, g(u_1)) : u_1 \in U_1\}$ of g is a submanifold of $E = E_1 \oplus E_2$ of dimension equal to that of E_1.

Proof. Identify E_2 with $\mathbf{R}^r$ by means of linear coordinates, and let $(g_1, \ldots, g_r)$ be the components of g. The graph of g is the set of points $(u, x_1, \ldots, x_r)$ of the open subset $U_1 \times \mathbf{R}^r$ of E such that $x_i - g_i(u) = 0$, $i = 1, \ldots, r$. If we put $f_i(u, x_1, \ldots, x_r) = x_i - g_i(u)$, this set is just the set of common zeros of the functions f_i. As the Df_i are linearly independent at each point, Theorem 9.1 gives the result.

We shall prove that each point m of a submanifold M has an open neighbourhood U such that $U \cap M$ is the graph of a map g as above. This gives a fourth characteristic of submanifolds of dimension d: *locally*, these are graphs of maps from open subsets of $\mathbf{R}^d$ to $\mathbf{R}^r$.

Proposition 9.1

Let M be a d-dimensional submanifold in $\mathbf{R}^n = \{(x_1, \ldots, x_n)\}$. Permuting $x_1, \ldots, x_n$ if necessary, every point $m \in M$ has a neighbourhood U such that $U \cap M$ is the graph of a map g of an open subset of $\mathbf{R}^d = \{(x_1, \ldots, x_k, 0, \ldots, 0)$ to:

$$\mathbf{R}^{n-d} = \{(0, \ldots, 0, x_{d+1}, \ldots, x_n)\}.$$

Proof. By Theorem 9.1, there is a neighbourhood U of m and $n-d$ functions $f_i: U \to \mathbf{R}$ such that $U \cap M$ is the set of common zeros of the f_i, the derivatives Df_i being linearly independent.

Let $f: U \to \mathbf{R}^{n-d}$ be the map with components f_i. The Jacobian matrix of f has an $(n-d) \times (n-d)$ minor with non-zero determinant. In other words, permuting the coordinates of $\mathbf{R}^n$ if necessary, we may write $\mathbf{R}^n = \{(u_1, u_2): u_1 \in \mathbf{R}^d,\ u_2 \in \mathbf{R}^{n-d}\}$, and so $D_2 f \in \mathscr{L}(\mathbf{R}^{n-d}, \mathbf{R}^{n-d})$ is an isomorphism. By the implicit function theorem (Corollary 3.2), there is a map $g: u_1 \mapsto u_2$ of an open subset U_1 of $\mathbf{R}^d$ to $\mathbf{R}^{n-d}$ such that $f(u_1, g(u_1)) = 0$. The set of zeros of f is thus the graph $\{(u_1, g(u_1)): u_1 \in U_1\}$ of g.

Example

The circle $S^1 = \{(x, y) \in \mathbf{R}^2: x^2 + y^2 = 1\}$ is locally (but not globally) the graph of differentiable maps of $\mathbf{R}$ to $\mathbf{R}$: S^1 is the union of the $U_i \cap S^1$, where $U_1 = \{|x| < 1,\ y > 0\}$, $U_2 = \{|x| < 1,\ y < 0\}$, $U_3 = \{x > 0,\ |y| < 1\}$, $U_4 = \{x < 0,\ |y| < 1\}$. The $U_i \cap S^1$, $i = 1, \ldots 4$, are the graphs of the maps:

$$x \mapsto y = \sqrt{(1-x^2)},\ x \mapsto y = -\sqrt{(1-x^2)},$$

$$y \mapsto x = \sqrt{(1-y^2)},\ y \mapsto x = -\sqrt{(1-y^2)}.$$

9.1.8 Parametrization

Let M be a d-dimensional submanifold of $\mathbf{R}^n$. We have just shown that each point $m \in M$ has an open neighbourhood U such that $U \cap M$ is the graph of a map g of an open subset V in $\mathbf{R}^d$ to $\mathbf{R}^{n-d}$. In other words, $U \cap M$ is the image of the map $p: x \in V \mapsto (x, g(x)) \in U \subset \mathbf{R}^d \times \mathbf{R}^{n-d} = \mathbf{R}^n$. The derivative $Dp(x) = (\mathrm{id}_{\mathbf{R}^d}, Dg(x))$ is clearly injective. Also, since p is a continuous map of V to U, it is also a continuous map of V on $V \cap M$ endowed with the induced topology. Lastly, the restriction q, to $M \cap U$, of the canonical projection $\mathbf{R}^n = \mathbf{R}^d \times \mathbf{R}^{n-d} \to \mathbf{R}^d$ on the first factor is continuous and $q \circ p = \mathrm{id}_V$, $p \circ q = \mathrm{id}_{U \cap M}$; thus p is a homeomorphism of V on $U \cap M$.

This shows that the condition stated in the following (fifth) characterization of submanifolds is necessary:

9.1.9 Further definition of submanifolds

Theorem 9.4

In order that M should be a C^k submanifold of dimension d of $\mathbf{R}^n$, it is necessary and sufficient that every point $m \in M$ should have an open neighbourhood U in $\mathbf{R}^n$ such that $U \cap M$ is the image of an open subset V of $\mathbf{R}^d$ under a C^k map $p: V \to U$ such that:

(a) $Dp(x)$ is injective for $x \in V$.

(b) p is a homeomorphism of V on $U \cap M$ endowed with the topology induced by $\mathbf{R}^n$.

We say that (V, p) is a parametrization (or a parametric representation) of $U \cap M$.

Proof. We simply have to show that the condition is sufficient. Set $a = p^{-1}(m)$. Since $Dp(a)$ is injective, it is of rank d and we can find d components of p whose derivatives are linearly independent at a. To fix ideas, suppose that these are the first d components. Then if $\pi_1 : \mathbf{R}^n = \mathbf{R}^d \times \mathbf{R}^{n-d} \to \mathbf{R}^d$ is the canonical projection on the first factor, $D(\pi_1 \circ p)(a) = \pi_1 \circ Dp(a)$ is an isomorphism and, by the inverse function theorem, $\pi_1 \circ p$ is a diffeomorphism of a neighbourhood of a on its image in $\mathbf{R}^d$. Denote by V this neighbourhood and let $U_1 = \pi_1 \circ p(V)$ be its image.

The function p maps V continuously to the open subset U of $\mathbf{R}^n$. By changing V if necessary, we may assume that U is the product of the open subset U_1 of $\mathbf{R}^d \times \{0\}$ by an open subset U_2 of $\{0\} \times \mathbf{R}^{n-d}$. Denote by $\pi_2 : \mathbf{R}^n = \mathbf{R}^d \times \mathbf{R}^{n-d} \to \mathbf{R}^{n-d} = \{0\} \times \mathbf{R}^{n-d}$ the canonical projection on the second factor. If $x \in V$ we have $p(x) = (\pi_1 \circ p(x), \pi_2 \circ p(x))$. Since $\pi_1 \circ p$ is a diffeomorphism of V on U_1, writing $y = \pi_1 \circ p(x)$ we may write $p(x) = (y, \pi_2 \circ p \circ [\pi_1 \circ p]^{-1}(y))$. This shows that $U \cap M$ is the graph of $\pi_2 \circ p \circ (\pi_1 \circ p]^{-1} : U_1 \to U_2$. The theorem then follows from Theorem 9.3. □

9.1.10 The torus

The torus T^2, obtained by rotating a circle of radius r about a line D in its plane at a distance $d > r$ from its centre, is a submanifold of $\mathbf{R}^3$.

To verify this, take the line D as the z-axis and let the centre of the circle be in the plane $z = 0$. The torus is the range of the map:

$$(\theta, \phi) \in \mathbf{R}^2 \rightsquigarrow (x = (d + r\cos\phi)\cos\theta,\ y = (d + r\cos\phi)\sin\theta,\ z = r\sin\phi) \in \mathbf{R}^3.$$

The restriction p of this map to the product $I \times J$ of two intervals of length less than 2π is a parametrization of T^2 (immediate verification).

9.1.11 The Möbius band

The range of the map $(\theta, t) \in \mathbf{R} \times]-1, 1[\to (x = (t\cos\theta + 2)\cos 2\theta,\ y = (t\cos\theta + 2)\sin 2\theta,\ z = t\sin\theta) \in \mathbf{R}^3$ is a submanifold of $\mathbf{R}^3$.

The restriction of this map to the product $I \times]-1, 1[$ of an interval of length less than π by $]-1, 1[$ is a parametrization (verify this).

Warning

A curve in $\mathbf{R}^n$—that is, a differentiable map f of an interval I in $\mathbf{R}$ to $\mathbf{R}^n$—does not necessarily have range which is a submanifold, even if f is a homeomorphism of I on $f(I)$. For example, $t \in \mathbf{R} \leadsto (t^2, t^3) \in \mathbf{R}^2$, because of the cusp at the origin.

9.2 THE TANGENT SPACE

Definition 9.1

Let M be a submanifold of a vector space E. A vector v of E is said to be tangent to M at m if it is the velocity vector at m of a curve traced on M and with origin m; that is, if there is an open interval I containing 0 and a differentiable map γ: $I \to E$ such that $\gamma(t) \in M$ for all $t \in I$, $\gamma(0) = m$ and $\gamma'(0) = v$.

The set of tangent vectors to M at m is called the tangent space to M at m and is written as T_mM.

Theorem 9.5

T_mM is a vector space of the same dimension as M.

Proof. By Definition 9.1, there is an open neighbourhood U of m and a diffeomorphism ϕ: $U \to E = \mathbf{R}^d \oplus \mathbf{R}^{n-d}$ such that $\phi(U \cap M) = \mathbf{R}^d \times \{0\}$. We may assume, without loss of generality, that $\phi(m) = 0$.

A curve γ traced on M and with origin m has image $\phi \circ \gamma$ which is a curve traced on the vector space $\mathbf{R}^d \times \{0\}$ and with origin $\phi(m) = 0$, and conversely.

It follows immediately that $D\phi(m)T_mM = \mathbf{R}^d \times \{0\}$, which shows that T_mM is the vector space $D\phi(m)^{-1}(\mathbf{R}^d \times \{0\})$ of dimension $d = \dim M$. □

Usually we identify the *vector* space T_mM with the affine space passing through m and parallel to T_mM (see Figure 9.3).

The different characterizations of submanifolds (five in §9.1) lead to as many characterizations of the tangent space. Here is that corresponding to the second definition.

Theorem 9.6

With the notation of Theorem 9.1, T_mM is the intersection of the kernels of the linear forms $Df_i(m)$.

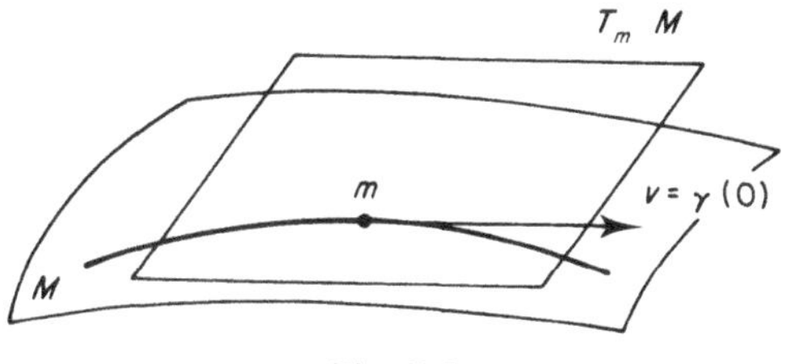

Fig. 9.3

Proof. This intersection is a vector space T of the same dimension d as T_mM. We thus have to prove that $T_mM \subset T$.

Let $\gamma: I \to E$ be a curve traced on M and with origin 0. Then $f_i[\gamma(t)] = 0$ for $t \in I$. Since the derivative is zero for $t = 0$, we obtain $Df_i(m)\gamma'(0) = 0$, which shows that $\gamma'(0) \in T$. □

Here is a characterization corresponding to the third definition of submanifolds.

Theorem 9.7

With the notation of Corollary 9.1, the tangent space at m to the submanifold of the level set $g^{-1}(b)$ is the kernel of $Dg(m)$.

Proof. This is similar to the proof of Theorem 9.6. □

Example

The sphere S^n in $\mathbf{R}^{n+1} = \{(x_1, \ldots, x_{n+1})\}$ is the submanifold of the level set $g^{-1}(1)$, where $g(x_1, \ldots, x_{n+1}) = x_1^2 + \ldots + x_{n+1}^2$.

A point $v \in T_mS^n$ is characterized by $Dg(m)v = 0$, that is $\langle m, v\rangle = 0$ (where $\langle , \rangle$ is the usual scalar product on $\mathbf{R}^{n+1}$). We recover the fact that T_mS^n is orthogonal to the 'radius' vector m of the sphere.

9.2.1 The graph point of view

Theorem 9.8

With the notation of Theorem 9.3, the tangent space at $m = (u, g(u))$ to the graph of g is the graph of $Dg(u)$.

Proof. Denote by $\pi_1 : E_1 \oplus E_2 \to E_1$ and $\pi_2 : E_1 \oplus E_2 \to E_2$ the canonical projections. The fact that a curve $\gamma: I \to E = E_1 \oplus E_2$ is traced on $M = \{(x, g(x)) \in E_1 \times E_2\}$ is expressed by writing $\pi_2 \circ \gamma(t) = g[\pi_1 \circ \gamma(t)]$ Differentiate this and use the fact that $\gamma(0) = (u, g(u))$. We obtain $\pi_2 \circ \gamma'(0) = Dg(u) \circ \pi_1 \circ \gamma'(0)$, which means that $\gamma'(0) = (\pi_1 \circ \gamma'(0), \pi_2 \circ \gamma'(0))$ belongs to the graph of $Dg(u)$ (and conversely).

9.2.2 The parametrization point of view

Theorem 9.9

We keep the notation of §9.1.8. Let (V, p) be a parametrization of $U \cap M$ and let $a \in V$. Then the tangent space to M at $m = p(a)$ is $Dp(a)\mathbf{R}^d$.

Proof. Let $s: I \to V$ be a differentiable curve with origin a. Then $p \circ s$ is a curve traced on M and with origin $p(a) = m$. Its velocity vector $D(p \circ s(0) = Dp(a)s'(0)$ describes a vector space T of dimension d when $s'(0)$ describes $\mathbf{R}^d$, as $Dp(a)$ is injective. Thus $T = Dp(a)\mathbf{R}^d$ is contained in T_mM and these two spaces are of dimension d: hence they coincide. □

Examples

(a) We have seen (Consequences in §9.1.1) that an open subset M of E is a submanifold of E of the same dimension as E. Its tangent space at $m \in M$ is simply E. For, with the notation of Theorem 9.5, we may choose $U = M$ and $\phi = \mathrm{id}_U$; then $T_mM = D\phi(m)^{-1}E = E$.

Thus the tangent space at an arbitrary point of the open submanifold GL(n) of End(E) = $\mathbf{R}^n$ (see §9.1.4), is simply End(E). We shall investigate whether this is so for several subgroups of GL(n) which are also submanifolds of End(E).

(b) We have seen (§9.1.5) that the unimodular group SL(n) of endomorphisms of $\mathbf{R}^n$ with determinant 1 is a submanifold of dimension $n^2 - 1$ of End($\mathbf{R}^n$). Let us find its tangent space at $I = \mathrm{id}_{\mathbf{R}^n}$.

Let N be the subspace of End($\mathbf{R}^n$) formed by the traceless endomorphisms u. With 'exp' denoting the exponential map, consider the curve γ: $t \in \mathbf{R} \rightsquigarrow \exp(tu) \in \mathrm{GL}(n)$, $u \in N$. This is a curve with origin I which, by Theorem 5.5, is traced on SL(n). By Theorem 5.6 its velocity vector at I is $\gamma'(0) = u$. The space $T_I\mathrm{SL}(n)$ thus contains N. Since $\dim T_I\mathrm{SL}(n) = \dim \mathrm{SL}(n) = n^2 - 1 = \dim N$, we see that $T_I\mathrm{SL}(n) = N$.

Later we shall learn how to determine the tangent space to SL(n) at an arbitrary point.

(c) We have seen §9.1.6 that the orthogonal group O(n) is a submanifold of End($\mathbf{R}^n$) of dimension $n(n-1)/2$. Let us find its tangent space at $I = \mathrm{id}_{\mathbf{R}^n}$.

Let A be the subspace of End($\mathbf{R}^n$) formed by the antisymmetric endomorphisms a: ${}^t a = -a$ (that is, $\langle a(x), x\rangle = 0$ for all $x \in \mathbf{R}^n$, where $\langle , \rangle$ is the usual scalar product). Consider the curve γ: $t \in \mathbf{R} \rightsquigarrow \exp(ta)$, where $a \in A$. If $F(t)$ denotes the scalar square of $\exp(ta)x$, where $x \in \mathbf{R}^n$, Leibniz' rule and Theorem 5.6 imply that $F'(t) = 2\langle a \exp(ta)x, \exp(ta)x\rangle$, which is zero since a is antisymmetric. Hence $F(t) = F(0)$, which shows that the curve γ of origin I is straced on O(n). By Theorem 5.6, $\gamma'(0) = a$. The space $T_I\mathrm{O}(n)$ thus contains A. As $\dim T_I\mathrm{O}(n) = \dim \mathrm{O}(n) = n(n-1)/2 = \dim A$, we see that $T_I\mathrm{O}(n) = A$.

(d) Endow $\mathbf{R}^{2n} = \{(q, p) : a \in \mathbf{R}^n, p \in \mathbf{R}^n\}$ with the symplectic form ω defined by $\omega[(q, p), (q', p')] = \langle q, p'\rangle - \langle q', p]$ (see Theorem 6.6).

The reader should prove that the subgroup Sp(n) of GL($2n$), formed by the operators s such that $\omega[s(\,), s(\,)'] = \omega[(\,), (\,)']$, is a submanifold of End($\mathbf{R}^{2n}$). He or she should also prove that $T_I\mathrm{Sp}(n)$ is made up of infinitesimal symplectic maps; that is, those $\sigma \in \mathrm{End}(\mathbf{R}^{2n})$ such that $\omega[\sigma(\,), (\,)'] + \omega[(\,), \sigma(\,)'] = 0$ for $(\,), (\,)' \in \mathbf{R}^{2n}$.

9.3 DIFFERENTIABLE MAPS

Definition 9.2

Let M, M' be submanifolds of vector spaces E, E' respectively, and let f: $M \to M'$. We say that f is of class C^k at $m \in M$ if there is an open neighbourhood U of m and a map $\hat{f}$: $U \to E'$ of class C^k such that $\hat{f}|_{U \cap M} = f_{U \cap M}$.

If f is of class C^k at each point of M, we say that f is of class C^k.

If M is an open subset of E, we may take $\hat{f} = f$ and recover the second definition.

9.3.1 Tangent linear maps

Theorem 9.10

With the preceding notation, $D\hat{f}(m)$ maps T_mM to $T_{f(m)}M'$, and its restriction to T_mM depends only on f and not on the choice of $\hat{f}$.

Proof. Let γ: $I \to U \cap M$ be a curve with origin m and traced on M. Then $\hat{f} \circ \gamma$ is a curve traced on M' and with origin $\hat{f}(m) = f(m)$. As $\hat{f} \circ \gamma(t) = f \circ \gamma(t)$ for $t \in I$, we have $D\hat{f}(m)\gamma'(0) = (\hat{f} \circ \gamma)'(0) = (f \circ \gamma)'(0)$, which shows that $D\hat{f}(m)\gamma'(0) \in t_{f(m)}M'$ does not depend on $\hat{f}$.

The restriction of $D\hat{f}(m)$ to T_mM is therefore the unique linear map $T_mf : T_mM \to T_{f(m)}M'$ such that $T_mf \,.\, \gamma'(0) = (f \circ \gamma)'(0)$ for every differentiable curve γ traced on M and with origin m. We call it the *tangent linear map of f at m*.

Example

If M is an open subset of E, then T_mf is simply $Df(m)$, regarded as a linear map of E to the subspace $Df(m)E$ of E'.

We may also interpret T_mf as the map $(m, \gamma'(0)) \rightsquigarrow (f(m), Df(m) \,.\, \gamma'(0))$ which we introduced in §1.2.6 as the tangent map.

As in §1.2.6 we have the following result, the *chain rule*, the proof of which is immediate.

Theorem 9.11 (Composition of maps)

If f: $M \to M'$ and g: $M' \to M''$ are C^k maps of C^k submanifolds, then $g \circ f$: $M \to M''$ is a C^k map and $T_m(g \circ f) = T_{f(m)}g \circ T_mf$. □

9.3.2 Diffeomorphisms of submanifolds

We say that f: $M \to M'$ is a diffeomorphism of the submanifold M on the submanifold M' if f is bijective and if f and f^{-1} are differentiable.

We say that f is a local diffeomorphism at $m \in M$ if there is an open neighbourhood U_M of m in M (for the topology induced by E) and an open neighbourhood $V_{M'}$ of $f(m)$ in M' (for the topology induced by E') such that the restriction of f to U_M is a diffeomorphism of U_M on $V_{M'}$.

In this case, $T_m f$ is bijective, for $f^{-1} \circ f = \mathrm{id}_{U_M}$ implies that $T_{f(m)} f^{-1} \circ T_m f = \mathrm{id}_{T_m}$.

Examples

(a) If (V, p) is a parametrization of an open subset $U_M = U \cap M$ of M, then p is a diffeomorphism of the open submanifold V of $\mathbf{R}^d$ on U_M.

(b) We make the identification $\mathrm{End}(\mathbf{R}^n) = \mathbf{R}^{n^2}$. If $u \in \mathrm{GL}(n)$, the map $L_u : \mathrm{End}(\mathbf{R}^n) \to \mathrm{End}(\mathbf{R}^n)$ defined by $L_u(g) = u \circ g$ is of class C^∞. We call it left-translation by u. It has an inverse, namely $L_{u^{-1}}$. It is therefore a diffeomorphism.

We may define in the same way a right-translation $R_u : g \mapsto g \circ u$.

Suppose that $u \in \mathrm{SL}(n)$. Clearly L_u leaves $\mathrm{SL}(n)$ invariant, since it is a group. Thus the restriction of L_u to $\mathrm{SL}(n)$ is a diffeomorphism of $\mathrm{SL}(n)$ on $\mathrm{SL}(n)$. This diffeomorphism therefore maps the tangent space N at $I = \mathrm{id}_{\mathbf{R}^n}$ on to the tangent space at u; in other words, $T_u \,\mathrm{SL}(n) = u \circ N$.

There is a similar result for $\mathrm{O}(n)$: $T_u \mathrm{O}(n) = u \circ A$.

9.3.3 The inverse function theorem for submanifolds

Theorem 9.12

With the previous notation, if $T_m f$ is bijective, f is a local diffeomorphism.

Proof. Choose parametrizations (p, V) and (p', V') of M and M' in neighbourhoods of m and $f(m)$ respectively. We may assume that $f \circ p\,(V) \subset p'(V')$.

Let $a = p^{-1}(m)$. Then by Theorem 9.11, $T_a(p'^{-1} \circ f \circ p) = T_{f(m)} p'^{-1} \circ T_m f \circ T_a p$ is bijective, as it is the composition of bijections. The inverse function theorem for vector spaces shows that there is an open neighbourhood of a in V (again denoted by V) and an open neighbourhood of $p'^{-1}(f(m))$ in V' (again denoted by V') such that $p'^{-1} \circ f \circ p : V \to V'$ is a diffeomorphism. Hence the composition of diffeomorphisms $f = p' \circ (p'^{-1} \circ f \circ p) \circ p^{-1}$ is a diffeomorphism. □

Notes

This chapter is merely an introduction to more complete treatments, such as that by P. Malliavin. The reader will also find, in the book by J. Milnor, direct applications of what is given above. In particular there is a geometrical proof of the fundamental theorem of algebra: every non-constant polynomial P: $C \to C$ has at least one root.

An application of Theorem 9.12 will be found in Appendix H.

10

Calculus of variations

This chapter is devoted to the study of the extrema of real-valued functions defined on a normed vector space E. Particular attention is paid to spaces E whose elements are curves in a vector space. The conditions necessary for an extremum are then expressed by a differential equation, the Euler–Lagrange equation.

Usually these questions originate in concrete situations (mechanics, physics) in which some symmetry is present. The invariance of the Euler–Lagrange equation under these symmetries is reflected in the existence of first integrals: this is the content of Noether's theorem.

10.1 FREE EXTREMA AND RESTRICTED EXTREMA

Definitions 10.1

Let A be a subset of a normed v.s. E, and let f be a real-valued function defined on A. We say that f has a *local minimum* at $a \in A$ if there is an open neighbourhood U of a in E such that $f(a) \leqslant f(x)$ for all $x \in U \cap A$.

In the same way a *local maximum* is defined. Moreover, f has a local maximum at a if, and only if, $-f$ has a local minimum at a.

We say that f has a *local extremum* at a if it has a local minimum or maximum at a.

The following result has been proved in §3.2.1.

Theorem 10.1

A differentiable function $f\colon A \to \mathbf{R}$ has a local extremum at an interior point a of A only if $Df(a) = 0$.

Remarks

(a) The criterion $Df(a) = 0$ is necessary but is not sufficient. For example, if $f\colon \mathbf{R} \to \mathbf{R}$ is defined by $f(t) = t^3$, then $Df(0) = 0$; but 0 is not a local extremum.

Again, if f: $\mathbf{R}^2 \to \mathbf{R}$ is defined by $f(x, y) = x^2 - y^2$, then $Df(0,0) = 0$; however, f takes positive and negative values in any neighbourhood of $(0,0)$

(b) If we use the condition $Df(a) = 0$ to determine the extrema of f on A, and if A is not open, we have to examine separately the values taken by f on the boundary of A. But this can be all of A as, for example, when A is a surface in $\mathbf{R}^3$. We shall see how to deal with this if the boundary of A is 'smooth enough'—more precisely, if it is a submanifold of E.

10.1.1 Restricted extremum

Let M be a differentiable submanifold, U an open subset of E which intersects M, and f a real-valued function defined on U. We say that f has a (local) restricted extremum at $m \in U \cap M$ if the restriction of f to $U \cap M$ has a local extremum at m.

Example

If M is a vector subspace of E, and if f is differentiable, then $f|_M$ is differentiable and $D(f|_M)(m) = Df(m)|_M$ (see §1.2.4).

The condition for a restricted extremum is thus expressed by $Df(m)|_M = 0$. This clearly does not imply that $Df(m) = 0$.

Suppose in addition that E is a euclidean space, with scalar product $\langle , \rangle$. The condition $Df(m)v = 0$ for all $v \in M$ may be written as $\langle \operatorname{grad} f(m), v \rangle = 0$. In other words, the gradient of f at m is orthogonal to M.

We shall now generalize this result.

Theorem 10.2

With the above notation, in order that the differentiable function f should have a restricted extremum at m on the submanifold M, it is necessary that the restriction of $Df(m)$ to the tangent space $T_m M$ be zero.

Proof. Let $v \in T_m M$, and let γ: $I \to E$ be a differentiable curve traced on M, with origin m and such that $\gamma'(0) = v$. Clearly $(f|_{M \cap U}) \circ \gamma(t) = f \circ \gamma(t)$ for $t \in I$, and this function of t has a local extremum at $t = 0$. By the chain rule (Theorem 9.11) and Theorem 10.1, we thus have $T_m(f|_{U \cap M})v = T_m f \circ v = Df(m)v = (\mathrm{d}/\mathrm{d}t)[f \circ \gamma](0) = 0$. □

To each of the characterizations of the tangent space corresponds a version of this result. We give the most useful, and to do that, begin with a lemma.

Lemma 10.1

Let a, $b_1, \ldots, b_k$ be linear functionals on an n-dimensional vector space E. Suppose that $b_1, \ldots, b_k$ are linearly independent and that the intersection of their kernels are contained in that of a. Then a is a linear combination of the b_i.

Proof. Complete the b_i in such a way as to form a basis $b_1, \dots, b_n$ of E^*, and let $e_1, \dots, e_n$ be the dual basis. Then there are constants c_i such that $a = \sum cb_i$. Also the intersection of the kernels of $b_1, \dots, b_k$ is the space generated by $e_{k+1}, \dots, e_n$. Thus $0 = a(e_r) = \sum c_i b_i(e_r) = c_r$ if $r = k+1, \dots, n$. Hence $a = c_1 b_1 + \dots + c_k b_k$. □

10.1.2 The Lagrange multiplier theorem

Theorem 10.3

We use the second definition of submanifolds (§9.1.2). Let U be an open subset of $E = \mathbf{R}^n$, and let $g_1, \dots, g_k$ be real-valued C^1 functions defined on U, whose derivatives are linearly independent at each point of U. We know that the set of the common zeros of the g_i is a submanifold M of $\mathbf{R}^n$.

Let f: $U \to \mathbf{R}$ be differentiable. Then f has a restricted extremum at $m \in U$ on M only if there are constants c_i such that $Df(m) = c_1 Dg_1(m) + \dots + c_k Dg_k(m)$.

Proof. By Theorem 9.6, $T_m M$ is the intersection of the kernels of the $Dg_i(m)$. By Theorem 10.2, $T_m M$ is contained in the kernel of $Df(m)$. It is now enough to use Lemma 10.1.

The extremum of f on M is sometimes called the extremum of f subject to the constraints $g_1 = \dots = g_k = 0$ (whence the name restricted extremum).

To find the point m, we may solve the system $D_j f(m) = \sum c D_j g_i(m)$, $j = 1, \dots, n$, to which we add $g_1(m) = \dots = g_k(m) = 0$. This is a system of $k+n$ equations in $k+n$ unknowns $c_1, \dots, c_k$, $m = (m_1, \dots, m_n)$. The auxiliary unknowns c_i are called *Lagrange multipliers.*

10.1.3 Two applications

(a) Given a surface M in $\mathbf{R}^3$, with equation $g(x, y, z) = 0$, and a point $A \in \mathbf{R}^3$, the points $m \in M$ whose distance from A is an extremum are given by the extrema of $f(m) = \langle m - A, m - A \rangle$ restricted by $g(m) = 0$.

Now $Df(m) = cDg(m)$ may be written as $2(m - A) = c$ grad $g(m)$. Geometrically, this means that the normal to M at the desired point m passes through A. But not all these points necessarily give an extremum: the sphere with centre A passing through m does not necessarily remain on the same side of M in a neighbourhood of m.

(b) Let u be a symmetric endomorphism of a euclidean space $E = \mathbf{R}^n$: $\langle u(x), y \rangle = \langle x, u(y) \rangle$ for all $x, y \in E$. We shall prove that E has a basis of eigenvectors of u.

The function f: $E \to \mathbf{R}$ defined by $f(x) = \langle u(x), x \rangle$ is differentiable. It thus attains its maximum on the unit sphere S (which is compact) at a point e_1. To find e_1, let us look for the restricted extrema of f constrained by $g(x) = \langle x, x\rangle = 1$. By Leibniz' rule, $Df(x)h = \langle u(x), h\rangle + \langle u(h), x\rangle = 2\langle u(x), h\rangle$ and $Dg(x)h = 2\langle x, h\rangle$ for $h \in E$. The relation $Df(e_1) = cDg(e_1)$ thus becomes $u(e_1) = ce_1$, where $f(e_1) = \langle u(e_1), e_1\rangle = c$. A point e_1 of $S = g^{-1}(1)$ at which f is maximum is therefore an eigenvector of u with maximum eigenvalue.

Now if $\langle y, e_1\rangle = 0$, we have $\langle u(y), e_1\rangle = \langle y, u(e_1)\rangle = c\langle y, e_1\rangle = 0$. The space E' orthogonal to e_1 is thus invariant under u, and the restriction of u to E' is again a symmetric endomorphism. Since $\dim E' = \dim E - 1$, we see, by induction on the dimension of E, that E has a basis of eigenvectors of u which are pairwise orthogonal.

10.2 SECOND-ORDER CONDITIONS FOR AN EXTREMUM

Let us return to a real-valued differentiable function f defined on an open subset U of a normed vector space E. We have seen that if f has an extremum at a, then $Df(a) = 0$. We shall obtain a supplementary necessary condition by assuming that f is twice differentiable.

Theorem 10.4

Let $f\colon U \to \mathbf{R}$ be twice differentiable at $m \in U$. If f has a local minimum at m, then $D^2f(m)(X, X) \geqslant 0$ for all $X \in E$.

Proof. Since $Df(m) = 0$, Taylor's formula gives $f(m+h) - f(m) = \frac{1}{2} D^2f(m)(h, h) + r(h)$, where $\lim_{h\to 0} r(h)/\| h \|^2 = 0$. As f has a local minimum at m, we have $f(m+h) - f(m) \geqslant 0$ for $\| h \|$ small enough, say for $\| h \| < \varepsilon$. Thus $D^2f(m)(h, h) + 2r(h) \geqslant 0$ for $\| h \| < \varepsilon$. Let $X \in E$ and take $t \in R$ so that $t \neq 0$, $\| tX \| < \varepsilon$. Since $D^2f(m)$ is bilinear, we obtain $t^2D^2f(m)(X, X) + 2r(tX) \geqslant 0$. Divide by t^2 and let t tend to zero. It follows that $D^2f(m)(X, X) \geqslant 0$.

If f has a local maximum at m, clearly $D^2f(m)(X, X) \leqslant 0$ for all $x \in E$.

Remarks

(a) The conditions $Df(m) = 0$ and $D^2f(m)(X, X) \geqslant 0$ for all $x \in E$ are not sufficient for f to have a local minimum at m. For example, if $f\colon \mathbf{R}^2 \to \mathbf{R}$ is defined by $f(x, y) = x^2 - 3y^3$, then $Df(0,0) = 0$, $D^2f(0,0)(X, Y) = 2X^2 \geqslant 0$. However, f takes positive and negative values in any neighbourhood of $(0, 0)$

(b) Suppose that f is three times differentiable. If $Df(m) = 0$, $D^2f(m) = 0$ but $D^3f(m) \neq 0$, then f cannot have a local extremum at m; for Taylor's formula gives $f(m + tX) - f(m) = (1/6)t^3D^3f(m)(X, X, X) + r(tX)$, where $\| r(tX) \|/t^3 \to 0$ as $t \to 0$. This shows that the left-hand side behaves (up to a sign) like t^3 near m; it therefore cannot be of constant sign.

10.2.1 A sufficient condition for a local minimum

Theorem 10.5

Let f be a C^3 real-valued function defined on an open subset U of a *Hilbert* space E. Suppose that $m \in U$ is a non-degenerate critical point of f and that $D^2f(m)(X,X) > 0$ for all $X \in E \setminus \{0\}$. Then f has a strict local minimum at m; that is, $f(m) < f(x)$ for all x near enough to m and distinct from m.

Proof. By the Morse–Palais theorem (Lemma 8.1), there is a local C^1-diffeomorphism ϕ on a neighbourhood of m such that $\phi(m) = m$ and $f(m+h) - f(m) = \frac{1}{2}D^2f(m)(\phi(h), \phi(h))$. The theorem follows.

The hypothesis here may be weakened. We have the following result.

Theorem 10.6

Let U be an open subset of a Banach space E, and let $f\colon U \to \mathbf{R}$ be twice differentiable at $m \in U$. IF m is a non-degenerate critical point of f, and if $D^2f(m)(X,X) > 0$ for all $X \in E \setminus \{0\}$, then f has a strict minimum at m.

Lemma 10.1

There is a constant $c > 0$ such that $D^2f(m)(X,X) \geqslant c \,\| X \|^2$ for all $X \in E$.

Proof. Put $L = D^2f(m)$. Since L is non-degenerate, the map $x \in E \rightsquigarrow L(x, .) \in E^*$ is an isomorphism. If M is the norm of the inverse isomorphism, then $\| x \|_E \leqslant M \| L(x, .) \|_{E^*}$. But $\| L(x, .) \|_{E^*} = \sup_{\| y \| = 1} | L(x, y) |$; thus, given x, there is $y \in E$ with norm 1 such that $2 \| L(x, y) \| \geqslant \| L(x, .) \|_{E^*}$. Thus:

$$\| x \| \leqslant 2M \,| L(x, y) | . \tag{10.1}$$

Since by Schwarz' theorem $L = D^2f(m)$ is symmetric, we have:

$$0 < L(x + ty, x + ty) = t^2L(y, y) + 2tL(x, y) + L(x, x) \text{ for } t \in \mathbf{R}.$$

The right-hand side is thus a quadratic form in t whose discriminant is negative; if $\| L \|$ is the norm of the continuous bilinear functional L (see Theorem A.7) we then have:

$$| L(x, y) |^2 \leqslant L(x, x)L(y, y) \leqslant \| L \| \, L(x, x).$$

With equation (10.1) this implies that $\| x \|^2 \leqslant 4M^2 \| L \| \, L(x, x)$. With $c^{-1} = 4M^2 \| L \|$ it follows that $c \| x \|^2 \leqslant L(x, x)$.

Proof of theorem 10.6 Taylor's formula and Lemma 10.1 imply that

$$f(m + h) - f(m) = \tfrac{1}{2}D^2f(m)(h, h) + r(h) \geqslant (c/2) \| h \|^2 + r(h)$$

where $r(h) = o(\| h \|^2)$. To the number $c/2$ corresponds $\delta > 0$ such that $\| h \| < \delta$ implies $| r(h) | \leqslant (c/4) \| h \|^2$. If $\| h \| < \delta$ we thus have $f(m+h) - f(m) \geqslant (x/4) \| h \|^2$. □

Remarks

(a) Consider a twice differentiable function f: $\mathbf{R} \to \mathbf{R}$ which has *only one critical point* a, which is non-degenerate and such that $f''(a) > 0$. Thereom 10.6 shows that a is a local minimum. In fact, it is a *global minimum*. For suppose that there was a point b, distinct from a, such that $f(b) \leqslant f(a)$. By the mean-value theorem, there would be a point c between a and b such that $f'(c) = 0$. This contradicts the uniqueness of the critical point a.

(b) The argument in (a) applies only to functions of a real variable. However, the result has the following generalization. Let f be a C^3 real-valued function defined on a *Hilbert* space E. Suppose that f has only one critical point m, that this critical point is non-degenerate, and that $D^2f(m)(h, h) > 0$ for all $h \in E$. Then $f(m) < f(x)$ for all $x \in E$ distinct from m.

Here is an idea of the proof. The integral curves of the vector field grad f all emanate from the point m. They cover E, and the restriction of f to any one of them is an increasing function, starting from a.

In what follows we are going to specialize the space E and the function f.

10.3 SPACES OF CURVES, AND THE EULER–LAGRANGE EQUATIONS

10.3.1 The space of C^1 curves

Let V be a normed vector space on $\mathbf{R}$, and let $I = [a, b]$, $a < b$. A C^1 map γ: $I \to V$ is called a C^1 curve, parametrized by $t \in I$, in the space V.

The set $C^1(I; V)$ of all such curves is clearly a v.s. on $\mathbf{R}$: if $k \in \mathbf{R}$ and if $\gamma_1, \gamma_2 \in C^1(I; V)$, we define C^1 curves $k\gamma_1$ and $\gamma_1 + \gamma_2$ by setting $(k\gamma_1)(t) = k\gamma_1(t)$, $(\gamma_1 + \gamma_2)(t) = \gamma_1(t) + \gamma_2(t)$.

The real vector space $C^1(I; V)$ has a natural norm. For since I is compact and γ and γ' are continuous, $\| \gamma(t) \|$ and $\| \gamma'(t) \|$ are bounded on I. We may thus set $\| \gamma \|_{C^1} = \sup_{t \in I} \| \gamma(t) \| + \sup_{t \in I} \| \gamma'(t) \|$. It is easy to check that $\| \gamma \|_{C^1}$ is a norm; we call it the C^1 norm.

Suppose that V is a Banach space. An obvious adaptation of the proof given in §A.1.3 (Example (c)) shows that $C^1(I; V)$ is complete with respect to the C^1 norm; it is therefore a Banach space.

10.3.2 Functions of curves

There are many ways in which we may make a real number correspond to a C^1 curve γ. Suppose that V is a euclidean space; we can, for example, attach to γ its euclidean length $\int_a^b \| \gamma'(t) \| \, dt$. We can also interpret $\gamma(t)$ as the position in V at time t of a particle of mass $+1$ and attach to it the action $\frac{1}{2}\int_a^b \| \gamma'(t) \|^2 \, dt$.

These examples are included in the following scheme, to which we shall confine ourselves.

Let L: $\mathbf{R} \oplus V \oplus V \to \mathbf{R}$ be a C^1 function (called the *lagrangian*). Given $\gamma \in C^1(I; V)$, $t \rightsquigarrow L(t, \gamma(t), \gamma'(t))$ is a continuous function. It is thus integrable, and we may define a function $\hat{L}: C^1(I; V) \to \mathbf{R}$ by setting:

$$\hat{L}(\gamma) = \int_a^b L(t, \gamma(t), \gamma'(t)) \, dt. \tag{10.2}$$

We shall prove that $\hat{L}$ is of class C^1 and shall calculate its derivative. We begin with a lemma which is interesting in its own right.

10.3.3 Differentiation under the integral

Lemma 10.2

Let U be an open subset of a normed v.s. E, let $I = [a, b]$, $a < b$, and let f be a continuous map of $I \times U$ to a normed v.s. F. Suppose that $D_2 f$ exists and is continuous; put $g(u) = \int_a^b f(t, u) \, dt$. Then g is of class C^1 on U and $Dg(u) = \int_a^b D_2 f(t, u) \, dt$.

Proof. Let $u \in U$. Choosing, if necessary, a small enough open neighbourhood of u, again denoted by U, we may assume that $D_2 f$ is bounded on $I \times U$. Let l: $E \to F$ be the continuous linear map $h \in E \rightsquigarrow \int_a^b D_2 f(t, u) h \, dt$. By the fundamental theorem of integral calculus:

$$\begin{aligned} g(u+h) - g(u) - l(u) &= \int_a^b [f(t, u+h) - f(t, u) - D_2 f(t, u) h] \, dt \\ &= \int_a^b \left[\int_0^1 D_2 f(t, u+sh) h \, ds - D_2 f(t, u) h \right] dt \\ &= \int_a^b \left\{ \int_0^1 [D_2 f(t, u+sh) - D_2 f(t, u)] h \, ds \right\} dt. \end{aligned}$$

It follows that:

$$\| g(u+h) - g(u) - l(h) \| \leqslant (b - a \| h \| \sup \| D_2 f(t, u+sh) - D_2 f(t, u) \|$$

where the supremum is taken over all s, t with $0 \leqslant s \leqslant 1$ and $a \leqslant t \leqslant b$. However $D_2 f$ is uniformly continuous on the compact set $I \times \{u\}$. Hence, given any $\varepsilon > 0$, there exists $\delta > 0$ such that $\| h \| < \delta$ implies that $\sup \| \quad \| < \varepsilon$. This shows that $l = Dg(u)$.

As $u \rightsquigarrow \int_a^b D_2 f(t, u) \, dt$ is continuous, g is of class C^1. □

Theorem 10.7

If the lagrangian L is of class C^1, then the map $\hat{L}: C^1(I; V) \to \mathbf{R}$ defined by equation (10.2) is of class C^1, and for all $h \in C^1(I; V)$ we have $D\hat{L}(\gamma)h = \int_a^b [D_2L() . h(t) + D_3L(\ ) . h'(t)]\, dt$, where we have abbreviated $(t, \gamma(t), \gamma'(t))$ by ( ).

Proof. We shall apply Lemma 10.2 with $U = E = C^1(I; V)$, $F = \mathbf{R}$ and $f(t, \gamma) = L(t, \gamma(t), \gamma'(t))$.

The map f is continuous as it is the composition of obviously continuous maps:

$$I \times C^1(I; V) \xrightarrow{M} I \oplus V \oplus V \xrightarrow{L} \mathbf{R},$$

$$(t, \gamma) \rightsquigarrow (t, \gamma(t), \gamma'(t)) \rightsquigarrow L(t, \gamma(t), \gamma'(t)).$$

We shall prove that $D_2 f$ exists and is continuous. Since L is of class C^1, it is enough to show that D_2M exists and is continuous. Let $\gamma,\ h \in C^1(I; V)$. Then $M(t, \gamma + h) - M(t, \gamma) = (0, h(t), h'(t))$. But the map $h \in C^1(I; V) \rightsquigarrow (0, h(t), h'(t)) \in I \oplus V \oplus V$, which is plainly linear, is continuous since $\| (0, h(t), h'(t)) \| = \| h(t) \| + \| h'(t) \| \leqslant \| h \|_{C^1}$. Thus D_2M exists and $D_2M(t, \gamma)h = (0, h(t), h'(t))$, which does not depend on γ and which is continuous in t.

Applying the chain rule to $f = L \circ M$, with $h \in C^1(I; V)$, we obtain:

$$D_2 f(t, \gamma)h = DL(M(t, \gamma)) . D_2M(t, \gamma)h = DL(t, \gamma(t), \gamma'(t)) . (0, h(t),\ h'(t))$$

$$D_2 f(t, \gamma)h = D_2L(\ )h(t) + D_3L(\ )h'(t) \tag{10.3}$$

where $(\ ) = (t, \gamma(t), \gamma'(t))$.

Since L is of class C^1, we see *a posteriori* that $D_2(t, \gamma)$ depends continuously on (t, γ).

With our choice of f, the hypotheses of Lemma 10.2 are satisfied and we obtain $D\hat{L}(\gamma)h = \int_a^b D_2 f(t, \gamma)h\, dt$ which, taking account of equation (10.3), proves Theorem 10.7 □

10.3.4 A restricted extremum problem

Let us retain the above notation. We intend to find the extrema of the function $\hat{L}: C^1(I; V) \to \mathbf{R}$, i.e. to find the curves γ which make $\hat{L}(\gamma) = \int_a^b L(t, \gamma(t), \gamma'(t))\, dt$ a minimum (or maximum).

In practice we meet several problems of the following type. Find the shortest path from A to B in the euclidean space V. This amounts to finding the C^1 curves γ with origin A and end B which make $\int_a^b \| \gamma'(t) \|\, dt$ a minimum.

More generally, given two points A, B in the space V in which the curves are traced, find a curve $\gamma \in C^1(I; V)$ with origin $\gamma(a) = A$ and end $\gamma(b) = B$ which makes $\hat{L}(\gamma)$ an extremum.

We are led to look for the extrema of the restriction of $\hat{L}$ to the subset

$\Gamma(A,B)=\{\gamma\in C^1(I;V):\gamma(a)=A,\ \gamma(b)=B\}$. It is clear that this subset is a closed affine subspace of $C^1(I;V)$: if $0\leqslant u\leqslant 1$ and if $\gamma_1,\gamma_2\in\Gamma(A,B)$, then $u\gamma_1+(1-u)\gamma_2\in\Gamma(A,B)$; and if $\gamma_n\in\Gamma(A,B)$ converges to $\gamma\in C^1(I;V)$ in the sense of the C^1 norm, then $\gamma(a)=\lim\gamma_n(a)=A$ and $\gamma(b)=\lim\gamma_n(b)=B$.

It can also be seen that $\Gamma(0,,)$ is a vector subspace and that, if γ_0 is a fixed element of $\Gamma(A,B)$—for example the segment $t\rightsquigarrow[(A-B)t+(aB-bA)]/(a-b)$—then $\Gamma(A,B)$ is obtainable from $\Gamma(0,0)$ by the translation $\gamma\rightsquigarrow\gamma+\gamma_0$.

By the trace theorem (§1.2.4), the restriction of $\hat{L}$ to $\Gamma(0,0)$ (thus to $\Gamma(A,B)$) is of class C^1 and the derivative of this restriction is the restriction to $\Gamma(0,\ 0)$ of $D\hat{L}$. It is important to understand that the derivative of the restriction of $\hat{L}$ to $\Gamma(A,B)$ is a continuous linear map which operates on $\Gamma(0,0)$: if $\gamma,\gamma+h\in(A,B)$ their difference $h\in\Gamma(0,0)$, by the observation made above. It follows that:

$$\hat{L}\,|\,_{\Gamma(A,B)}(\gamma+h)-\hat{L}\,|\,_{\Gamma(A,B)}(\gamma)=D\hat{L}(\gamma)h+o(h)$$

With the help of Theorems 10.1 and 10.7, we have thus proved the following result.

Theorem 10.8

The map $\hat{L}:\Gamma(A,B)\to\mathbf{R}$ has an extremum at γ only if:

$$\int_a^b[D_2L(t,\gamma(t),\gamma'(t))h(t)+D_3L(\ )h'(t)]\ dt=0$$

for all $h\in\Gamma(0,0)$, that is, for all $h\in C^1(I;V)$ such that $h(a)=h(b)=0$. □

Such a curve γ is usually referred to as *an extremal of the lagrangian L with fixed extremities A and B*. This terminology should not give rise to errors: the condition $D\hat{L}(\gamma)=0$ is necessary for the extremum, but is not sufficient, in general.

To determine the extremals, we shall give a more manageable form of the last theorem. To do this, we limit ourselves to curves γ traced on a *Hibert space V*. This enables us to identify V with its dual V^* by means of the map $x\in V\rightsquigarrow\langle x,\cdot\rangle\in V^*$.

10.3.5 Du Bois Reymond's lemma

Lemma 10.3

Let B: $I=[a,b]\to V^*$ be a continuous function with values in the dual of a Hilbert space V. In order that $\int_a^b B(t)h'(t)\ dt=0$ for all $h\in C^1(I;V)$ satisfying $h(a)=h(b)=0$, it is necessary and sufficient that B should be constant.

Proof. The condition is plainly sufficient. To show that it is necessary, note that, if $C: I \to V^*$ is a constant, we have:

$$\int_a^b [B(t) - C]h'(t)\,dt = 0.$$

Choose $C = (1/(b-a)) \int_a^b B(t)\,dt$, identify V with its dual, and take:

$$h(t) = \int_a^t [B(s) - C]\,ds.$$

Then $h(a) = h(b) = 0$; h is differentiable and $h'(t) = B(t) - C$ is continuous. We must therefore have:

$$0 = \int_a^b [B(t) - C]h'(t)\,dt = \int_a^b \| B(t) - C \|^2\,dt.$$

This implies that $B(t) = C$ for all $t \in I$. □

Corollary 10.1

Let A: $I = [a, b] \to V^*$ and B: $I \to V^*$ be continuous. In order that $\int_a^b [A(t)h(t) + B(t)h'(t)]\,dt = 0$ for all $h \in C^1(I; V)$ such that $h(a) = h(b) = 0$, it is necessary and sufficient that B should be differentiable and that $B' = A$.

Proof. The condition is sufficient, for $B'h + Bh' = \langle B, h\rangle'$, and thus:

$$\int_a^b [Ah + Bh']\,dt = \int_a^b \langle B, h\rangle'\,dt = \langle B(b), h(b)\rangle - \langle B(a), h(a)\rangle = 0.$$

To show that it is necessary, set $C(t) = \int_a^t A(s)\,ds$. Then $C'(t) = A(t)$, and so $Ch' + Ah = \langle C, h'\rangle + \langle C', h\rangle = \langle C, h\rangle'$. Integrate and use the facts that $h(a) = h(b) = 0$:

$$0 = \langle C(b), h(b)\rangle - \langle C(a), h(a)\rangle = \int_a^b \langle C, h\rangle'\,dt$$

$$= \int_a^b [C(t)h'(t) + A(t)h(t)]\,dt.$$

By hypothesis, $\int_a^b [B(t) - C]h'(t)\,dt = 0$ and, by Lemma 10.3, $B(t) = C(t) + \text{constant}$. Hence B' exists and equals $C' = A$. □

Theorem 10.9

Let V be a Hilbert space. In order that $\gamma \in C^1(I; V)$ should be an extremal of the lagrangian L with fixed extremities $\gamma(a) = A$ and $\gamma(b) = B$, it is necessary and sufficient that $t \mapsto D_3L(t, \gamma(t), \gamma'(t))$ be differentiable and that:

$$\frac{d}{dt} D_3L(t, \gamma(t), \gamma'(t)) = D_2L(t, \gamma(t), \gamma'(t)) \text{ for all } t \in I. \qquad (10.4)$$

Proof. This is an immediate consequence of Theorem 10.8 and Corollary 10.1, in which we take $A = D_2L$ and $B = D_3L$. Equation (10.4) is known as the Euler–Lagrange equation. It determines the extremals. Note that the values of γ at a and b do not occur in it.

Remark

In the book by H. Cartan (pp. 294–296) there is a proof of Du Bois Reymond's lemma which is valid not only in Hilbert spaces but also in real normed linear spaces.

10.3.6 Free extrema problems

Let us look for the extrema of $\hat{L}: C^1(I; V) \to \mathbf{R}$ without requiring that $\gamma(a) = A$ and $\gamma(b) = B$. If $\hat{L}$ has an extremum at γ, then $D\hat{L}(\gamma) = 0$. The restriction of $D\hat{L}(\gamma)$ to every affine subspace of $C^1(I; V)$ passing through γ is thus zero. In particular, let us take the affine subspace $\Gamma(A, B)$ and apply Theorem 10.9. We see that γ must also satisfy the Euler–Lagrange equations. This is, moreover, clear by Theorem 10.7: if $D\hat{L}(\gamma)h = 0$ for *all* $h \in C^1(I; V)$, it must certainly be so for all h satisfying $h(a) = A$, $h(b) = B$.

Conversely, if γ satisfies the Euler–Lagrange equations, we may ask whether $D\hat{L}(\gamma) = 0$. The reader should prove that for all $h \in C^1(I; V)$:

$$D\hat{L}(\gamma)h = D_3L(b, \gamma(b), \gamma'(b)) - D_3L(a, \gamma(a), \gamma'(a)),$$

from which the result follows.

Remark

To establish the Euler–Lagrange equation, we have assumed that the lagrangian L was defined on the whole space $I \oplus V \oplus V$. But it can happen that the images of the curves γ stay confined in a subset of V in which they are traced. It may also be that the velocity $\gamma'(t)$ remains between certain bounds. It is therefore desirable to restate Theorem 10.7 under the sole hypothesis that L is only defined on $I \times 0$, where 0 is an open subset of $V \oplus V$. We must thus study $\gamma \mapsto \hat{L}(\gamma)$ when $\gamma \in \Omega = \{\gamma \in C^1(I; V) : (\gamma(t), \gamma'(t)) \in 0\}$.

We shall prove that Ω is an *open subset* of $C^1(I; V)$. This will make it sensible to ask if $\gamma \in \Omega \mapsto \hat{L}(\gamma) \in \mathbf{R}$ is differentiable; and as differentiability is a local property, all the results of this section will hold, and in particular, Theorems 10.7 and 10.9.

Let $\gamma_0 \in \Omega$ and $\varepsilon > 0$. Let $\gamma \in C^1(I; V)$ be such that $\| \gamma - \gamma_0 \|_{C^1} < \varepsilon$. By the definition of the C^1 norm, we thus have $\| \gamma(t) - \gamma_0(t) \|_V < \varepsilon$ and $\| \gamma'(t) - \gamma_0'(t) \|_V < \varepsilon$ for all $t \in I$. By the definition of the norm on $V \oplus V$, it follows that:

$$\| (\gamma(t), \gamma'(t)) - (\gamma_0(t), \gamma_0'(t)) \|_{V \oplus V} = \| \gamma(t) - \gamma_0(t) \|_V + \| \gamma'(t) - \gamma_0'(t) \|_V < 2\varepsilon.$$

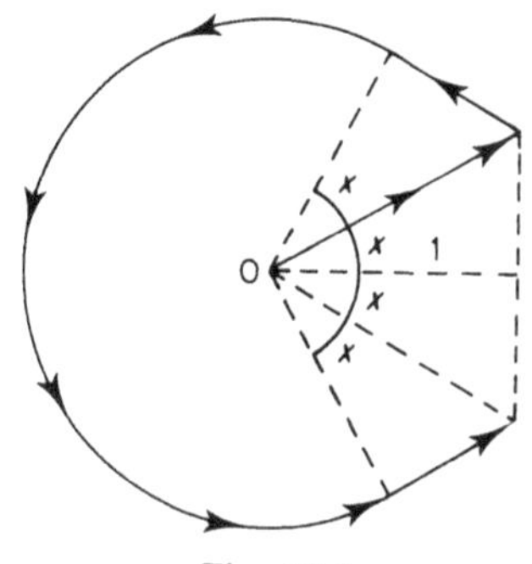

Fig. 10.1

Hence the set $K = \{(\gamma(t), \gamma'(t)) : t \in I\}$ is at a distance less than 2ε from the compact subset $\{(\gamma_0(t), \gamma_0'(t)) : t \in I\}$ of 0. As 0 is an open subset of $V \oplus V$, we may choose ε so small that $K \subset 0$.

If $\gamma_0 \in \Omega$, the ball in $C^1(I; V)$ with centre γ_0 and radius ε is thus contained in Ω, which is therefore an open set.

Remark

It should not be thought that an extremum problem, with differentiable data, necessarily has a differentiable solution. Here is a counter-example, put in the form of a mathematical puzzle.

A man, lost at sea, knows that he is at a distance r from a straight shoreline. But the fog is so thick that he does not know in what direction this beach is to be found. What is the path of minimum length he must follow to be certain of finding land? Put mathematically: find a plane curve with origin 0, of minimum length and which intersects every line in the plane situated at a distance r from the origin.

We can see in Figure 10.1 that the curve is not differentiable. Take $r = 1$. Its length, as a function of the angle x measured in radians, is $2\pi - 4x - 2\tan x + (\cos x)^{-1}$. We find that the value of x which minimizes this is about 7.28.

10.4 THE NATURE OF THE EULER–LAGRANGE EQUATION

Let us return to the lagrangian $L(t, q, v)$. We have seen that if $\gamma: I \to V$ is an extremal, then $t \in I \rightsquigarrow D_3L[t, \gamma(t), \gamma'(t)]$ is differentiable. As this is a function of t which depends on $\gamma(t)$ and $\gamma'(t)$, we would like to use the chain rule. To justify this, we have to introduce additional hypotheses, by assuming that D_3L and γ' are differentiable. The Euler–Lagrange equation may then be written as $D_{13}L(\ ) + D_{23}L(\ )\gamma'(t) + D_{33}(\ )\gamma''(t) = D_2L(\ )$. This is a second-order differential equation for γ.

Suppose also that $D_{33}L(\ ) \in \mathscr{L}(V^*; V)$ is invertible in the domain of $(\ ) = (t, \gamma(t), \gamma'(t))$. With $\gamma' = v$ the last equation may be written as a first-order differential system in (γ, v):

$$\gamma'(t) = v(t)$$
$$v'(t) = [D_{33}L(\;)]^{-1}[D_2L(\;) - D_{13}L(\;) - D_{23}L(\;)v(t)] \tag{10.5}$$

where $(\;) = (t, \gamma(t), \gamma'(t))$. This leads to the following notion.

10.4.1 Regular lagrangians

Let 0 be an open subset of $V \oplus V$. We say that a lagrangian $L: I \times 0 \to \mathbf{R}$ is regular if it is of class C^2 and if $D_{33}L(t, q, v) \in \mathscr{L}(V^*; V)$ is an isomorphism for all $t \in I$, $(q, v) \in 0$.

Theorem 10.10

The extremals $\gamma: I \to V$ of a regular lagrangian L are of class C^2. Given any $t_0 \in I$ and $(q, v_0) \in 0 \subset V \oplus V$, there is a maximal interval $J \subset I$ containing t_0 and a unique extremal $\gamma: J \to V$ such that $\gamma(t_0) = q_0$, $\gamma'(t_0) = v_0$.

Proof. Set $D_3L(t, q, v) = p$, where $(q, v) \in 0$; and look for v, being given t, q and p. Since $D_{33}L(\;)$ is invertible, the implicit function theorem (Theorem 3.2) shows that, locally, there is a function G, of the same C^1 class as D_3L, such that $v = G(t, q, p)$.

We now write the *Euler–Lagrange* equation, using the facts that $v = \gamma'$ is the derivative of $q = \gamma$, and that $(\mathrm{d}/\mathrm{d}t)D_3L(\;) = D_2L(\;)$. We obtain the first-order differential system:

$$\frac{\mathrm{d}q}{\mathrm{d}t} = G(t, q, p),$$

$$\frac{\mathrm{d}p}{\mathrm{d}t} = D_2L[t, q, G(t, q, p)].$$

Since the right-hand sides are of class C^1, the solutions are of class C^1. In particular, $\gamma' = \mathrm{d}q/\mathrm{d}t$ is of class C^1; that is, γ is of class C^2. As $p = D_3L(t, q, v)$ is of class C^1 and γ is of class C^2, it is now legitimate to write the equations in the form of equations (10.5).

The second part of the theorem is an immediate consequence of the existence and uniqueness theorem (Theorem 7.4). □

Remarks

(a) It is not necessary that the lagrangian be regular for the conclusions of Theorem 10.10 to hold. Suppose, for example, that $L(t, q, v)$ does not depend on v. Then D_2L is identically zero and Euler-Lagrange equation reduces to $D_2L[t, \gamma(t)] = 0$, which defines $\gamma(t)$ implicity if $D_{22}L(\;) \in \mathscr{L}(V^*; V)$ is invertible.

Here is another example. Suppose that $L(t, q, v)$ does not depend on q. Then D_2L is identically zero and the Euler-Lagrange equation shows that $D_3L[t, \gamma'(t)]$ is constant. It may happen that we can find $\gamma'(t)$ from this equation even if $D_{33}L(\;)$ is not invertible. This is so if, for example, V is the euclidean space $\mathbf{R}^n$ and $L(t,q,v) = \| v \|$. We obtain $L^{-1}v = D_3 L(\,) =$ constant; that is, $\gamma'(t) =$ constant. Thus the extremals $\gamma(t)$ are lines described at constant velocity. We would expect this, for the extremal problem consists of looking for the curves of minimum length $\int_a^b \| \gamma'(t) \| \,\mathrm{d}t$. However, the lagrangian is not regular, for we see that $D_{33} L(v) v = 0$.

(b) Theorem 10.10 solves the Cauchy problem for the Euler–Lagrange equation when the lagrangian is regular: $t_0 \in I$ being fixed, through a point q_0 of V passes a unique extremal with given velocity $\gamma'(t_0) = v_0$.

However, this theorem does not solve the problem that we posed initially: find an extremal γ given $\gamma(a) = A$ and $\gamma(b) = B$.

Even if this last problem were solved, it would still remain to check that γ really was an extremum of L. This problem lies outside the scope of this book.

10.4.2 Example of a regular lagrangian

Let N particles, denoted by 1 to N and with respective masses $m_i > 0$, be situated in euclidean space $\mathbf{R}^3$. Denote by $q_i(t)$ and $v_i(t)$ their positions and velocities respectively at time t.

Suppose they are subjected to forces derivable from a C^1 potential $U(q_1, \ldots, q_N)$. Newton's equations of motion may be written as $m_i(\mathrm{d}v_i/\mathrm{d}t) = -\partial U/\partial q_i$, where $\partial U/\partial q_i = D_i U.1$, $D_i U$ being the partial derivative of U with respect to $q_i \in \mathbf{R}^3$.

We may represent these N points by a single point $q = (q_1, \ldots, q_N)$ of $\mathbf{R}^{3N}$, the velocity of which is $v = (v_1, \ldots, v_N) \in \mathbf{R}^{3N}$. Introduce the kinetic energy $T = \frac{1}{2}\sum m_i \| v_i \|^2$ and the lagrangian L: $\mathbf{R} \oplus \mathbf{R}^{3N} \oplus \mathbf{R}^{3N} \to \mathbf{R}$ defined by $L(q,v) = T - U$. Then Newton's equations coincide with the Euler–Lagrange equations; for, with obvious notation:

$$\frac{\partial L}{\partial v_i} = \frac{\partial T}{\partial v_i} = m_i v_i \text{ and } \frac{\partial L}{\partial q_i} = -\frac{\partial U}{\partial q_i};$$

thus

$$0 = m_i \frac{\mathrm{d}q_i}{\mathrm{d}t} + \frac{\partial U}{\partial q_i} = \frac{\mathrm{d}}{\mathrm{d}t}\left(\frac{\partial L}{\partial v_i}\right) - \frac{\partial L}{\partial q_i}.$$

This is *Hamilton's principle of least action*: the motion of the system is given by the extremals of the action integral $\int (T - U)\,\mathrm{d}t$.

The lagrangian is regular, for $D_{33}L(\,)$ has matrix with components:

$$\frac{\partial^2 L}{\partial v_i \partial v_j} = \begin{cases} 0 \text{ if } i \neq j \\ m_i \text{ if } i = j \end{cases}$$

If we put $D_3 L = p$, we find that the components $p_1, \ldots, p_N$ of p are $p_i = m_i v_i$, which is simply the momentum of the ith particle. The function G of Theorem 10.10 is very simple: $v_i = p_i/m_i$. The first-order system written in Theorem 10.10 is here:

$$\frac{\mathrm{d}q_i}{\mathrm{d}t} = \frac{p_i}{m_i}, \quad \frac{\mathrm{d}p_i}{\mathrm{d}t} = -\frac{\partial U}{\partial q_i}.$$

With $H(q, p) = \frac{1}{2}\Sigma \| p_i \|^2/m_i + U(q)$, this becomes

$$\frac{dq_i}{dt} = \frac{\partial H}{\partial p_i}, \quad \frac{dp_i}{dt} = -\frac{\partial H}{\partial q_i}$$

This is the hamiltonian form of the equations of motion (see §7.1.5).

Suppose that the system reduces to one particle of mass $+1$ and that it is not subjected to any external force. The Euler–Lagrange equation (or Newton's equation) shows that the velocity of the particle remains constant. The extremals are straight lines described at constant speed.

Here the lines are not simply extremals of the length $\int_a^b \| \gamma'(t) \| \, dt$, but they are also extremals of the action $\int_a^b \| \gamma'(t) \|^2 \, dt$. The following example is a generalization of this result.

Example

Let $L\colon \mathbf{R} \oplus V \oplus V \to \mathbf{R}$ be a lagrangian independent of $t \in \mathbf{R}$ and such that, for each $Q \in V$, $v \in V \rightsquigarrow L(q, v)$ is a positive non-degenerate quadratic form. In other words, there is a bilinear, symmetric, non-degenerate form $b(q)$ such that $L(q, v) = b(q)(v, v) > 0$ if $v \neq 0$, the function $q \rightsquigarrow b(q)$ being of class C^1.

We propose to show that every extremal γ of $\int_a^b L \, dt$ is an extremal of $\int_a^b \sqrt{L} \, dt$. Let us begin by proving that if γ is an extremal of L, $L[\gamma(t), \gamma'(t)]$ does not depend on t. For $dL/dt = D_2L \, . \, \gamma' + D_3L \, . \, \gamma'' = [D_2L - (d/dt)D_3L]\gamma' + (d/dt)(D_3L \, . \, \gamma')$. But the first term on the right-hand side is zero, by the Euler–Lagrange equation; and Leibniz' rule shows that $D_3L \cdot v = D_3[b(q)(v, v)]v = 2b(q)(v, v) = 2L$. Hence $0 = (d/dt)L$.

The last relation and the Euler–Lagrange equation for L show that:

$$\frac{d}{dt} D_3(\sqrt{L}) - D_2(\sqrt{L}) = \frac{d}{dt}\left[\tfrac{1}{2}L^{-1/2}D_3L\right] - \tfrac{1}{2}L^{-1/2}D_2L$$

$$= -\frac{1}{4}L^{-3/4}\frac{dL}{dt} + \tfrac{1}{2}L^{-1/2}\left[\frac{d}{dt}D_3L - D_2L\right] = 0$$

This is the Euler–Lagrange equation for the lagrangian $\sqrt{L}$; γ is thus an extremal of $\sqrt{L}$ (see H. Cartan, pp. 303–306).

10.5 EFFECT OF A DIFFERENTIABLE MAP

We return to the space $C^1(I; V)$ of C^1 curves $\gamma\colon I \to V$. If $\phi\colon V \to V$ is a map of class C^k, $k \geqslant 1$, the image $\phi \circ \gamma$ of a C^1 curve is a C^1 curve; thus ϕ induces a map $\Phi\colon C^1(I; V) \to C^1(I; V)$ defined by $(\Phi\gamma)(t) = \phi[\gamma(t)]$.

Since $C^1(I; V)$ is a space normed by the C^1 norm, we may ask whether Φ is differentiable.

Proposition 10.1

If ϕ is of class C^2, then Φ is of class C^1, and the value at $h \in C^1(I; V)$ of its derivative at $\gamma \in C^1(I; V)$ is given by $D\Phi(\gamma)h: t \rightsquigarrow D\phi[\gamma(t)] \,.\, h(t)$ for all $t \in I$.

Proof.
Step 1: Fix $\gamma \in C^1(I; V)$. If $h \in C^1(I; V)$, then since $D\phi$, γ and h are of class C^1, the map $t \in I \rightsquigarrow D\phi[\gamma(t)]h(t) \in V$ is of class C^1. Hence this defines a map L of $C^1(I; V)$ into itself: $L(h)$: $t \in I \rightsquigarrow D\phi[\gamma(t)]h(t)$, which is plainly linear. We now prove that it is continuous.

We must find a constant A such that $\| L(h) \|_{C^1} \leqslant A \| h \|_{C^1}$ for all h. If we put $A_1 = \sup_{t \in I} \| D\phi[\gamma(t)] \|$, then:

$$\| D\phi[\gamma(t)]h(t) \| \leqslant \| D\phi[\gamma(t)] \| \; \| h(t) \| \leqslant A_1 \| h \|_{C^1}.$$

Also $D^2\phi[\gamma(t)](\gamma'(t), .)$ is a continuous linear functional on V which depends continuously on t. Denoting by A_2 the supremum of its norm when $t \in I$, we have:

$$\begin{aligned} \left\| \frac{\mathrm{d}}{\mathrm{d}t} \{D\phi[\gamma(t)]h(t)\} \right\| &= \| D^2\phi[\gamma(t)](\gamma'(t), h(t)) + D\phi[\gamma(t)]h'(t) \| \\ &\leqslant \| D^2\phi[\gamma(t)](\gamma'(t), .) \| \; \| h(t) \| \\ &\quad + \| D\phi[\gamma(t)] \| \; \| h'(t) \| \leqslant (A_1 + A_2) \| h \|_{C^1}. \end{aligned}$$

To sum up, by the definition of the C^1 norm we have $\| L(\gamma)h \|_{C^1} \leqslant (2A_1 + A_2) \| h \|_{C^1}$.

Step 2: It remains to show that $\| \Phi(\gamma + h) - \Phi(\gamma) - L(\gamma)h \|_{C^1}$ divided by $\| h \|_{C^1}$ tends to zero with $\| h \|_{C^1}$. By the definition of the C^1 norm, we must prove that given any $\varepsilon > 0$, there exists $\delta > 0$ such that $\| h \|_{C^1} < \delta$ implies that $\| \Phi(\gamma + h)(t) - \Phi(\gamma)(t) - L(\gamma)h(t) \| = \| \phi[\gamma(t) + h(t)] - \phi[\gamma(t)] - D\phi[\gamma(t)]h6t) \| \leqslant \varepsilon \| h \|_{C^1}$ for all $t \in I$, and an analogous inequality for the derivative with respect to t of $\Phi(\gamma + h) - \Phi(\gamma) - L(\gamma)h$.

We shall simply prove the first of these; the second may be established in the same way. Let γ, $h \in C^1(I; V)$. Since γ and h are continuous, the set $K = \{\gamma(t) + sh(t): t \in I, 0 \leqslant s \leqslant 1\}$ is a compact subset of V which contains the images of γ and $\gamma + h$. Since $D^2\phi$ is continuous, $\| D^2\phi[\gamma(t) + sh(t)] \|$ is bounded on K by a number A. Apply Taylor's formula (Theorem 4.7):

$$\| \phi[\gamma(t) + h(t)] - \phi[\gamma(t)] - D\phi[\gamma(t)]h(t) \| \leqslant A \| h(t) \|^2 \leqslant A(\| h \|_{C^1})^2$$

If $\delta < \varepsilon A^{-1}$ we have the result. □

Corollary 10.2

If ϕ is a C^2 diffeomorphism, then Φ is a C^1 diffeomorphism. □

Now let us consider a C^1 lagrangian L: $I \oplus V \oplus V \to \mathbf{R}$ and go back to the search for the extrema of $\hat{L}(\gamma) = \int_a^b L[t, \gamma(t), \gamma'(t)] \,\mathrm{d}t$.

Theorem 10.11

The property that a curve $\gamma \in C^1(I;\, V)$ has of being extremal does not depend on the coordinate system chosen in the space V in which it is traced.

Proof. Identify V with $\mathbf{R}^n$. Let (U, ϕ) be a chart whose domain contains the image of γ; ϕ is a diffeomorphism of U on $\phi(U)$. By Corollary 10.2, ϕ induces a diffeomorphism Φ of the set of curves of $C^1(I;\, V)$ with images in U, to the set of curves of $C^1(I;\, \mathbf{R}^n)$ with images in $\phi(U)$. By the chain rule, writing $\hat{L} = (\hat{L} \circ \Phi^{-1}) \circ \Phi$, we have $D\hat{L}(\gamma) = \mathrm{D}[\hat{L} \circ \Phi^{-1}](\Phi(\gamma)) \circ D\Phi(\gamma)$. Since $D\Phi(\gamma)$ is an isomorphism, this shows that $D\hat{L}(\gamma) = 0$ implies that $D[\hat{L} \circ \Phi^{-1}](\Phi(\gamma)) = 0$. Thus the image $\Phi(\gamma)$ of the extremal γ in the chart is the extremal of the expression $\hat{L} \circ \Phi^{-1}$ of $\hat{L}$ in the chart. □

This theorem is very important. In mechanics, for example, it enables us to write the equation of motion in an arbitrary coordinate system, by using Hamilton's principle of least action. We give only the following simple example.

Example

A particle of mass $m > 0$, situated in a plane $\mathbf{R}^2$, is subjected to a potential which depends only on the distance ρ of the particle from the origin.

Take polar coordinates (ρ, θ) with origin 0. The lagrangian $T - U$ introduced in §10.4.2 may be written as;

$$L(\rho, \theta, \rho', \theta') = \tfrac{1}{2} m(\rho'^2 + \rho^2 \theta'^2) - U(\rho).$$

The Euler–lagrange equation decomposes into two scalar equations:

$$\frac{\mathrm{d}}{\mathrm{d}t}\left(\frac{\partial L}{\partial \rho'}\right) = \frac{\partial L}{\partial \rho}, \frac{\mathrm{d}}{\mathrm{d}t}\left(\frac{\partial L}{\partial \theta'}\right) = \frac{\partial L}{\partial \theta}.$$

The first of these gives $m\rho'' = m\rho\theta'^2 - \mathrm{d}U/\mathrm{d}\rho$. As θ does not appear in L, the second gives $\partial L/\partial \theta' =$ constant; that is, $m\rho^2\theta' =$ constant. This is the law of conservation of angular momentum about 0, also called the 'law of areas'.

We shall see in §7 how such conservation laws may be included in a more general scheme.

10.6 INVARIANCE OF A LAGRANGIAN

Definition 10.2

Suppose that the lagrangian $L\colon I \oplus V \oplus V \to \mathbf{R}$ does not depend on $t \in I$. It is then a C^1 function defined on $V \oplus V$. In this case it is convenient to interpret $V \oplus V$ as the set $TV = \{q \in V, v \in V\}$ of pairs formed by a point $q \in V$ and by a vector v tangent to V at that point (we regard V as a submanifold of V).

The lagrangian L is then a C^1 function on TV. If $\phi: V \to V$ is a C^k diffeomorphism, it defines (§1.2.6) a tangent map $T\phi: TV \to TV$, by the rule $(q, r) \rightsquigarrow (\phi(q), D\phi(q)v)$, which is a C^{k-1} diffeomorphism.

We shall say that L is invariant under ϕ if $L = L \circ T\phi$; in other words, if:

$$L(q, v) = L[\phi(q), D\phi(q)v] \text{ for } q \in V, v \in V.$$

Example

Suppose that $V = \mathbf{R}^3 = \{(q_1, q_2, q_3)\}$. Let $U: \mathbf{R}^3 \to \mathbf{R}$ be a given C^1 function which does not depend on q_1, and consider the lagrangian:

$$L(q_1, q_2, q_3, v_1, v_2, v_3) = \tfrac{1}{2}(v_1^2 + v_2^2 + v_3^2) - U(q_2, q_3)$$

Then the translation $(q_1, q_2, q_3) \rightsquigarrow (q_1 + s, q_2, q_3)$ leaves L invariant.

Theorem 10.12

If $\gamma: I = [a, b] \to V$ is an extremal of L with ends $\gamma(a) = A$ and $\gamma(b) = B$, and if the diffeomorphism $\phi: V \to V$ leaves L invariant, then $\phi \circ \gamma$ is also an extremal of L, with endpoints $\phi(A)$ and $\phi(B)$.

Proof. With the notation of the last paragraph, invariance means that $\hat{L}(\Phi\gamma) = \hat{L}(\gamma)$. Thus $D\hat{L}(\Phi\gamma) \circ D\Phi(\gamma) = D\hat{L}(\gamma)$. If γ is an extremal, $D\hat{L}(\gamma) = 0$. Since $D\Phi(\gamma)$ is invertible, it follows that $D\hat{L}(\Phi\gamma) = 0$. This shows that $\phi \circ \gamma$ is an extremal.

Remark

Many authors regard this result as obvious. Here is the probable reason. Suppose that γ achieves the minimum of $\hat{L}(\gamma)$. Then $\phi \circ \gamma$ also achieves it, since $\hat{L}(\Phi\gamma) = \hat{L}(\gamma)$; $\phi \circ \gamma$ is thus an extremal. Unfortunately the condition $D\hat{L}(\gamma) = 0$ is not sufficient to guarantee that we have an extremal; it is merely necessary.

10.6.1 The Poincaré half-plane

Here is a spectacular application of Theorem 10.12 which shows that considerations of invariance can greatly simplify calculations.

Let the space V be the plane $\mathbf{R}^2 = \{(x, y)\}$, and let the lagrangian be the function $L(x, y, \dot{x}, \dot{y}) = (\dot{x}^2 + \dot{y}^2)/y^2$ defined on the product of the half-plane:

$$P = \{(x, y) : y > 0\} \text{ by } \mathbf{R}^2 = \{(\dot{x}, \dot{y})\}.$$

We shall look for those extremals γ whose images lie in P.

Step 1: It is clear that the lagrangian is regular. By Theorem 10.10, it follows that given $q_0 \in P$ and $v_0 \in \mathbf{R}^2$, there is a unique extremal γ such that $\gamma(0) = q_0$ and $\gamma'(0) = v_0$.

Step 2: Complexify P by associating the complex number $z = x + iy$ to the point (x, y). With an obvious abuse of notation, we may write $L(z, \dot{z}) = |\dot{z}|^2/[\mathrm{Im}(z)]^2$, where Im stands for 'imaginary part'.

An easy calculation shows that L is invariant under the symmetry s: $(x, y) \rightsquigarrow (-x, y)$, and under the homographies h: $z \rightsquigarrow (az+b)/(cz+d)$, where the real numbers a, b, c and d satisfy $ad - bc = 1$.

Step 3: Recall that these homographies leave P globally invariant and that they map the half-line $x = 0$, $y > 0$ into a half-line $x = x_0$, $y > 0$ or into a semicircle centred on the x-axis. Also recall that given $(q_0, v_0) \in TP$, where $L(q_0, v_0) = 1$, there is a homography h of the above type which sends the point $e = (1, 0)$ to q_0 and the vector $u = (0, 1)$ to v_0: $h(e) = q_0$, $Dh(e)u = v_0$.

Step 4: We look for the extremal γ such that $\gamma(0) = e$, $\gamma'(0) = u$ (it is unique by step 1). By step 2, $s \circ \gamma$ is also an extremal. As it satisfies the same initial conditions $s \circ \gamma(0) = e$, $(s \circ \gamma)'(0) = s(\gamma'(0)) = s(u) = u$ as γ, it must coincide with γ. Since γ is invariant under the symmetry s, its image is thus the half-line $x = 0$, $y > 0$.

Step 5: Let $q_0 \in P$ and $v_0 \in \mathbf{R}^2$ be such that $L(q_0, v_0) = 1$. By step 1, there is a unique extremal π satisfying $\pi(0) = q_0$, $\pi'(0) = v_0$.

Also (by step 3) there is homography h such that $h(e) = q_0$, $Dh(e)u = v_0$. By theorem 10.12 if γ is the extremal of step 4, $h \circ \gamma$ is also an extremal. As it satisfies the same initial conditions $(h \circ \gamma)(0) = q_0$ and $(h \circ \gamma)'(0) = Dh(e)u = v_0$ an π, we must have $\pi = h \circ \gamma$ (by uniqueness).

By step 3, it follows that the image of every extremal π is either a half-line $x = x_0$, $y > 0$ or a semicircle in P centred on the x-axis (see Figure 10.2).

Now consider any C^1 curve l: $t \in [a, b] \rightsquigarrow (x(t), y(t)) \in P$, and call the number $\int_a^b \sqrt{(x'^2 + y'^2)}/y \, dt$ its length in the sense of Lobatchevskii. We shall look for the curves l of minimum 'length' joining two given points of P. This leads us to find the extremals of the lagrangian $\sqrt{L} = \sqrt{(x'^2 + y'^2)}/y$. Combining the previous results and those of the example in §10.4.2, we see that these curves have images which are either half-lines $x = x_0$, $y > 0$ or semicircles in P centred on the x-axis.

If we call the distance (in the sense of lobatchevskian geometry) between two points A and B in P, the length of the extremal which joins them, these half-

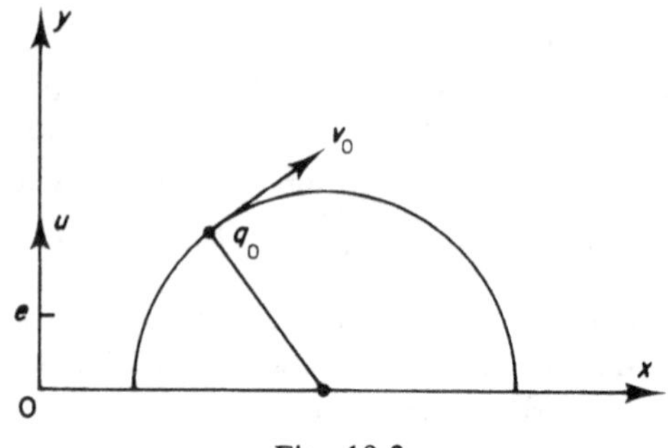

Fig. 10.2

lines and semicircles play the part of straight lines in euclidean geometry. It is easy to see that the symmetry s and the homographies h play the rôles of displacement in euclidean geometry.

P is called the Poincaré half-plane, after the mathematician who introduced the model of Lobatchevskii's geometry which we have just described.

10.7 EMMY NOETHER'S THEOREM

We shall see how the invariance of a lagrangian under a one-parameter group of diffeomorphisms produces functions which remain constant along extremals.

Definition 10.3

We say that a lagrangian $L: I \oplus V \oplus V \to \mathbf{R}$, which is independent of $t \in I$, is invariant under a one-parameter group of diffeomorphisms ϕ_s (with parameter s) of V if, for all s, L is invariant under ϕ_s.

Theorem 10.13

Let L be a lagrangian which is independent of t and invariant under a one-parameter group of diffeomorphisms ϕ_s of V. Then the function $\mathbf{J}: V \oplus V \to \mathbf{R}$, defined by $(q, v) \rightsquigarrow D_3L(q, v)[(\partial/\partial s)\phi_s(q)|_{s=0}] \in \mathbf{R}$, where D_3L is the partial derivative of L with respect to v, is constant along each extremal of L. We say that it is a first integral of the Euler–Lagrange equation.

Proof. Let $\gamma: I \to V$ be an extremal. By hypothesis, $L[\phi_s(\gamma(t)), (\mathrm{d}/\mathrm{d}t)(\phi_s \circ \gamma)(t)]$ does not depend on s. Since its derivative with respect to s is zero, we have:

$$D_2L[\ ]\,.\,\frac{\partial}{\partial s}[(\phi_s \circ \gamma)(t)] + D_3L[\ ]\,.\,\frac{\partial}{\partial s}\frac{\mathrm{d}}{\mathrm{d}t}(\phi_s \circ \gamma)(t) = 0.$$

But by Schwarz' theorem, $(\partial/\partial s)(\mathrm{d}/\mathrm{d}t) = (\mathrm{d}/\mathrm{d}t)(\partial/\partial s)$. Also, the Euler–Lagrange equation may be written as $D_2L = (\mathrm{d}/\mathrm{d}t)D_3L$. Hence:

$$\frac{\mathrm{d}}{\mathrm{d}t}D_3L[\ ]\,.\,\frac{\partial}{\partial s}[(\phi_s \circ \gamma)(t)] + D_3L[\ ]\,.\,\frac{\mathrm{d}}{\mathrm{d}t}\frac{\partial}{\partial s}(\phi_s \circ \gamma)(t) = 0.$$

that is:

$$\frac{\mathrm{d}}{\mathrm{d}t}\left\{D_3L[\ ]\,.\,\frac{\partial}{\partial s}[(\phi_s \circ \gamma)(t)]\right. = 0.$$

Set $s = 0$; with the notation of the theorem we have:

$$\frac{\mathrm{d}}{\mathrm{d}t}\left\{D_3L[\gamma(t), \gamma'(t)]\,.\,\frac{\partial}{\partial s}\phi_s \circ \gamma(t)|_{s=0}\right\} = 0. \qquad \square$$

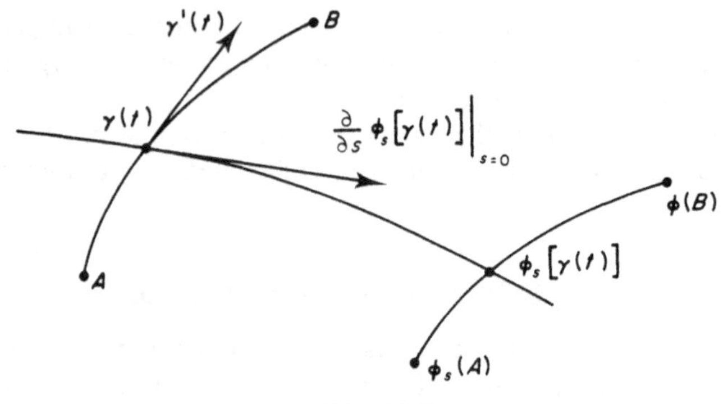

Fig. 10.3

Remark

$(\partial/\partial s)(\phi_s \circ \gamma)(t)\,|_{s=0}$ may be interpreted as the velocity vector at $s=0$ of the point $\gamma(t)$ describing its orbit under the action of the group ϕ_s (see Figure 10.3).

10.7.1 Invariance under translation

Suppose that L is invariant under a one-parameter group of translations with parameter s: $\phi_s(q) = q + sa$, where $a \in V$. Then $\phi_s[\gamma(t)] = \gamma(t) + sa$ and $(\partial/\partial s)\phi_s[\gamma(t)]\,|_{s=0} = a$.

The corresponding first integral is thus $D_3L[\gamma(t), \gamma'(t)]\,a$.

Suppose now, to consider an even more special case, that L is invariant under *every* translation. We may thus choose a arbitrarily and Noether's theorem means that the linear functional $D_3L[\gamma(t), \gamma'(t)]$ does not depend on t. This is easy to establish directly. Since L is invariant under all translations, $L(q, v)$ does not depend on q. Thus $D_2L = 0$, and the Euler–Lagrange equations $(\mathrm{d}/\mathrm{d}t)D_3L = D_2L$ show that $D_3L = \text{constant}$.

Now consider in §10.4.2 again. If the lagrangian $L = T - U$ is invariant under all translations, U is a constant.

By what has been shown above, $D_3L = D_3T = \text{constant}$. Thus $\sum m_i v_i = \text{constant}$, that is, the momentum is conserved in time.

10.7.2 Invariance under rotation

Let V be a euclidean space $\mathbf{R}^3$ oriented canonically. This will enable us to use the vector product $\wedge$.

Suppose that the lagrangian $L(q, v)$ is invariant under rotation around an axis passing through the origin $0 \in \mathbf{R}^3$. If ω is a vector along this axis, it is easy to check that $(\partial/\partial s)\phi_s(q)\,|_{s=0} = \omega \wedge q$. Noether's theorem thus can be written as $D_3L[\gamma(t), \gamma'(t)] \cdot (\omega \wedge q) = \text{constant}$.

Identifying a linear functional with a vector by means of the isomorphism $x \in V \mapsto \langle v, \cdot \rangle \in V^*$, this can be written as $\langle D_3L, \omega \wedge q \rangle = \text{constant}$. Taking account of the properties of the mixed product, $\langle q \wedge D_3L[\ ], \omega \rangle = \text{constant}$.

Again consider the example in §10.4.2. Invariance under rotation is represented by $\langle \sum m_i(q_i \wedge v_i), \omega \rangle = \text{constant}$. If, in particular, the potential U depends only on the distance from the origin, $L = T - U$ is invariant under *all* rotations about 0. We may thus choose ω arbitrarily, from which it turns out that the angular momentum $\sum m_i(q_i \wedge v_i)$ about 0 is constant.

10.7.3 Generalization of Noether's theorem

Suppose that $L: I \oplus V \oplus V \to \mathbf{R}$ depends on $t \in I$. We shall use a trick to bring about a reduction to the time-independent case.

Introduce space–time $\bar{V} = \mathbf{R} \oplus V = \{(t, q) : t \in \mathbf{R}, q \in V\}$, and let $\pi_2 : \bar{V} \to V$ be the second projection. If $\gamma: I \to V$ is a curve in V, it can be lifted up to a curve $\bar{\gamma} : t \in I \mapsto (t, \gamma(t)) \in \bar{V}$ and $\pi_2 \circ \bar{\gamma} = \gamma$.

Define a lagrangian L_1 on the tangent space $T\bar{V} = \bar{V} \oplus \bar{V} = \{(t, q, u, v\}$ by:

$$L_1(t, q, u, v) = L(t, q, v/u)u \quad \text{for} \quad u \neq 0.$$

If $\bar{\gamma} : s \in I \mapsto (t(s), \gamma(s)) \in \bar{V}$ is a C^1 curve on $\bar{V}$, then:

$$\hat{L}_1(\bar{\gamma}) = \int_a^b L_1\left[t(s), \gamma(s), \frac{dt}{ds}, \frac{d\gamma}{ds}\right] ds = \int_a^b L\left[t(s), \gamma(s), \frac{d\gamma}{ds}\Big/\frac{dt}{ds}\right] \frac{dt}{ds}\, ds$$

$$= \int_a^b L\left[t(s), \gamma(s), \frac{d\gamma}{dt}\right] dt = \hat{L}(\pi_2 \circ \bar{\gamma}).$$

It follows that $D\hat{L}_1(\bar{\gamma}) = D\hat{L}(\pi_2\bar{\gamma}) \circ \pi_2 = D\hat{L}(\gamma) \circ \pi_2$. Thus if $\bar{\gamma}$ is an extremal of L_1, its projection $\gamma = \pi_2 \circ \bar{\gamma}$ is an extremal of L and, conversely, the lifting $\bar{\gamma}$ of an extremal γ of L is an extremal of L_1 (taking as parameter $s = t$).

If ϕ_r is a one-parameter subgroup of diffeomorphisms of space–time $\bar{V}$ which leaves L_1 invariant, then by Noether's theorem there is a first integral $\mathbf{J}_1(t, q, u, v)$. If γ is an extremal of L, its lifting $\bar{\gamma}$, parametrized by $s = t$, is an extremal of L_1. Thus $\mathbf{J}_1(t(s), \gamma(s), (dt/ds), (d\gamma/ds)) = \mathbf{J}_1(t, \gamma(t), 1, (d\gamma/dt))$ does not depend on $s = t$: it is a first integral of L.

Example (conservation of energy)

If L does not depend on $t \in I$, then L_1 is invariant under the one-parameter group of time translations $\phi_s : (t, q) \mapsto (t + s, q)$. Thus $(\partial/\partial s)\phi_s(t, \gamma(t))\,|_{s=0} = (1, 0) \in \mathbf{R} \oplus V$.

Noether's theorem then implies that $D_3L_1(\ ).(1, 0) = \text{constant}$, where D_3 denotes the partial derivative with respect to $(u, v) \in \bar{V}$. This can also be written as $(\partial/\partial u)L_1(\ ) = \text{constant}$, or, taking account of the expression defining L_1, as $L(t, \gamma(t), \gamma'(t)) - D_3L(t, \gamma(t), \gamma'(t)).\gamma'(t) = \text{constant}$. This is what is called the first energy integral. The reason is that, if we again consider the example in §10.4.2 and the time-independent lagrangian $L(q, v) = T - U$, then the corresponding first integral is (up to a sign) $-L + D_3L \, . \, v = T + U$. This is the total energy.

Bibliography

The book by L.C. Young cannot be recommended too highly. Not only is a humourous gem found in each paragraph, but a detailed study of the calculus of variations is completed by that of its descendent optimal control. The reader who loves geometry will be enchanted by the last chapter of a book by P. Malliavin.

Appendix A

Banach spaces and multilinear maps

In all that follows the field K over which the vector spaces (v.s.) are constructed is the field of the real or complex numbers.

A.1. BANACH SPACES

A.1.1 Norms

A norm on a v.s. E is a function $\| \ \|: E \to \mathbf{R}$ such that $\|x\| \geqslant 0$, $\|x\| = 0$ if and only if $x = 0$, $\|x+y\| \leqslant \|x\| + \|y\|$, $\|kx\| = |k| \, \|x\|$ for all $k \in K$ and all $x, y \in E$. These properties imply that $| \, \|x\| - \|y\| \, | \leqslant \|x-y\|$.

A v.s. supplied with a norm is called a normed space.

Examples

(a) Each of the following maps is a norm on $\mathbf{R}^n$ (or $\mathbf{C}^n$):

$$(x_1, \ldots, x_n) \mapsto \begin{cases} (x_1^2 + \ldots + x_n^2)^{1/2} \\ |x_1| + \ldots + |x_n| \\ \sup\{|x_1|, \ldots, |x_n|\} \end{cases}$$

(b) Let $C_b^0(X)$ be the v.s. of functions on a topological space X, with values in K, which are continuous and bounded. Then

$$\|f\|_{C^0} = \sup_{x \in X} |f(x)|$$

is a norm, called the uniform convergence norm, or C^0 norm.

(c) Let $C^1([a,b])$ be the v.s. of functions f defined on an interval $[a,b]$, $a < b$, with real values and having a continuous derivative at each point. Then

$$\|f\|_{C^1} = \sup_{x \in [a,b]} |f(x)| + \sup_{x \in [a,b]} |f'(x)|$$

is a norm, called the C^1 norm. Note that $\|f\|_{C^1} = \|f\|_{C^0} + \|f'\|_{C^0}$.

A.1.2 The metric induced by a norm

Let E be a normed v.s., and put $d(x, y) = \| x - y \|$ for $x, y \in E$. Then d is a metric on E, which then becomes a metric space, and hence a topological space.

A.1.3 Banach spaces

These are normed v.s. which are complete with respect to the metrics induced by their norms.

Examples

(a) $\mathbf{R}^n$ (or $\mathbf{C}^n$) is a Banach space for each of the three norms above.
(b) $C_b^0(X)$ is a Banach space with the uniform convergence norm.
(c) Let us show that the normed space $C^1([a, b])$ of Example (c) above is a Banach space. Let (f_n) be a Cauchy sequence. Since $\| f \|_{C^1} = \| f \|_{C^0} + \| f' \|_{C^0}$, each of the sequences (f_n) and (f_n') is a Cauchy sequence in $C^0([a, b])$; as this space is complete, f_n (resp. f_n') converges uniformly to a function f (resp. g) in $C^0([a, b])$. Since

$$f_n(x) - f_n(a) = \int_a^x f_n'(t)\, \mathrm{d}t, \quad a \leqslant x \leqslant b$$

we have

$$\left| f(x) - f(a) - \int_a^x g(t)\, \mathrm{d}t \right| = \left| f(x) - f_n(x) + f_n(a) - f(a) + \int_a^x [f_n'(t) - g(t)]\, \mathrm{d}t \right|$$

$$\leqslant | f(x) - f_n(x) | + | f_n(a) - f(a) | + \int_a^x | f_n'(t) - g(t) |\, \mathrm{d}t$$

$$\leqslant 2 \| f - f_n \|_{C^0} + (b - a) \| f_n' - g \|_{C^0}$$

which tends to zero as $n \to \infty$. Thus f' exists and $f' = g$, for $f(x) = f(a) + \int_a^x g(t)\, \mathrm{d}t$.

It follows that (f_n) converges to f in the C^1 norm.

A.1.4 Direct sum of normed spaces.

Let $E_1, \ldots, E_n$ be normed spaces over the same field, with norms $\| \; \|_1, \ldots, \| \; \|_n$ respectively. It can be shown that the following are true.
(a) Their cartesian product becomes a v.s., written $E_1 \oplus \ldots \oplus E_n$ and called the direct sum of $E_1, \ldots, E_n$, under the operations

$$(x_1, \ldots, x_n) + (y_1, \ldots, y_n) = (x_1 + y_1, \ldots, x_n + y_n)$$

$$k(x_1, \ldots, x_n) = (kx_1, \ldots, kx_n), k \in K.$$

(b) $\|(x_1, \ldots, x_n)\| = \|x_1\|_1 + \ldots + \|x_n\|_n$ is a norm on $E_1 \oplus \ldots \oplus E_n$.
(c) $E_1 \oplus \ldots \oplus E_n$ is a Banach space if each of the spaces $E_1, \ldots, E_n$ is a Banach space.

A.2 CONTINUOUS LINEAR MAPS

Let E and F be normed spaces over the same field, and let $f: E \to F$ be a linear map.

Theorem A.1

The following conditions are equivalent:
(a) f is continuous at every point.
(b) f is continuous at the origin.
(c) $\sup_{\|x\| \leqslant 1} \|f(x)\| < \infty$ (in other words, the image of the unit ball in E is bounded.
(d) $\sup_{\|x\| = 1} \|f(x)\| < \infty$ (in other words, the image of the unit sphere in E is bounded.
(e) There is a constant M such that $\|f(x)\| \leqslant M\|x\|$ for all $x \in E$.

Proof. Clearly (a) $\Rightarrow$ (b), so let us show that (b) $\Rightarrow$ (c). There exists $\delta > 0$ such that $\|x\| \leqslant \delta \Rightarrow \|f(x)\| \leqslant 1$. Then $\|x\| \leqslant 1 \Rightarrow \|\delta x\| \leqslant \delta \Rightarrow \|f(\delta x)\| \leqslant 1 \Rightarrow \|f(x)\| \leqslant \delta^{-1}$.

Plainly (c) $\Rightarrow$ (d), so now we show that (d) $\Rightarrow$ (e). Let $M = \sup_{\|x\| = 1} \|f(x)\|$. If $x \neq 0$, $x/\|x\|$ has norm 1, and thus $\|f(x)\|/\|x\| \leqslant M$ and $\|f(x)\| \leqslant M\|x\|$; the inequality also holds if $x = 0$.

Finally, (e) $\Rightarrow$ (a) since $\|f(x) - f(y)\| = \|f(x - y)\| \leqslant M\|x - y\|$. □

A.2.1 Norm of a continuous linear map

The number $\sup_{\|x\| = 1} \|f(x)\|$ is called the norm of f and is written $\|f\|$. The proof that (d) $\Rightarrow$ (e) above shows that $\|f(x)\| \leqslant \|f\| . \|x\|$ for all $x \in E$.

We shall now prove that the continuous linear maps f of E to f form a v.s., denoted by $\mathscr{L}(E; F)$, and that $f \mapsto \|f\|$ is a norm (which justifies its name). Let $f, g \in \mathscr{L}(E; F)$. If $x \in E$, $\|x\| = 1$, then $\|(f + g)(x)\| = \|f(x) + g(x)\| \leqslant \|f(x)\| + \|g(x)\| \leqslant \|f\| + \|g\|$; thus $f + g$ is continuous and $\|f + g\| \leqslant \|f\| + \|g\|$. On the other hand, if $k \in K$, then $\|(kf)(x)\| = \|kf(x)\| = |k| . \|f(x)\|$, and so

$$\sup_{\|x\| - 1} \|(kf)(x)\| = |k| . \sup_{\|x\| - 1} \|f(x)\|$$

that is, $\|kf\| = |k| . \|f\|$. Lastly, $\|f\| = 0 \Rightarrow f(x) = 0$ for all $x \Rightarrow f = 0$.

Theorem A.2

The Banach space $\mathscr{L}(E; F)$: If F is a Banach space, then $\mathscr{L}(E; F)$ is complete.

Proof. Let (f_n) be a Cauchy sequence in $\mathscr{L}(E; F)$. Fix $x \in E$. The inequality $\| f_p(x) - f_q(x) \| \leqslant \| f_p - f_q \| . \| x \|$ shows that $(f_n(x))$ is a Cauchy sequence in f. It therefore has a limit, denoted by $f(x)$.

We next prove that f is linear:

$$f(x+y) = \lim_{n\to\infty} f_n(x+y) = \lim_{n\to\infty} [f_n(x) + f_n(y)] = \lim_{n\to\infty} f_n(x) + \lim_{n\to\infty} f_n(y)$$
$$= f(x) + f(y).$$

Similary $f(kx) = kf(x)$, $k \in K$.

Next we prove that f is continuous. Given any $\varepsilon > 0$, there exists on N such that $p, q \geqslant N \Rightarrow \| f_p - f_q \| \leqslant \varepsilon$; that is, $\| (f_p - f_q)(x) \| = \| f_p(x) - f_q(x) \| \leqslant \varepsilon$ if $\| x \| = 1$. Letting q tend to ∞, we obtain $\| f_p(x) - f(x) \| \leqslant \varepsilon$ for $\| x \| = 1$. From this we first deduce that $\| f(x) \| \leqslant \| f(x) - f_p(x) \| + \| f_p(x) \| \leqslant \varepsilon + \| f \|_p$ if $\| x \| = 1$; thus f is continuous. We also deduce that $\| f - f_p \| \leqslant \varepsilon$ if $p \geqslant N$; thus (f_p) converges to f. □

A.2.2 Norm of a product of continuous linear maps

Let E, F and G be three normed spaces over the same field, and let $u: E \to F$, $v: F \to G$ be continuous linear maps. Their composition $v \circ u: E \to G$ is plainly linear and continuous. Let $x \in E$, $\| x \| = 1$: then $\| v \circ u(x) \| = \| v[u(x)] \| \leqslant \| v \| \; \| u(x) \| \leqslant \| v \| \; \| u \|$. Thus $\| v \circ u \| \leqslant \| v \| \; \| u \|$.

A.2.3 Isomorphism of normed spaces

Let E and F be normed spaces over the same field. A map $f: E \to F$ is called an isomorphism (of normed spaces) if $f \in \mathscr{L}(E; F)$, f is a bijection and the inverse bijection f^{-1} (which is *ipso facto* linear) is continuous.

Theorem A.3

Let $f: E \to F$ be a *surjective* linear map. The following conditions are equivalent:

(a) f is an isomorphism.

(b) There are two numbers $m > 0$ and $M > 0$ such that, for all $x \in E$, $m \| x \| \leqslant \| f(x) \| \leqslant M \| x \|$.

Proof. We first show that (a) $\Rightarrow$ (*b*). Since f and f^{-1} are continuous, there are two constants $A > 0$ and $B > 0$ such that $\| f(x) \| \leqslant A \| x \|$ for all $x \in E$ and $\| f^{-1}(y) \| \leqslant B \| y \|$ for all $y \in f$. If $x \in E$, put $y = f(x)$, so that $x = f^{-1}(y)$. Then $\| x \| \leqslant B \| f(x) \|$ and we have $B^{-1} \| x \| \leqslant \| f(x) \| \leqslant A \| x \|$.

Now we prove that (b) $\Rightarrow$ (a). The inequality $\| f(x) \| \leqslant M \| x \|$ shows that f is continuous. The inequality $m \| x \| \leqslant \| f(x) \|$, $m > 0$, shows that $\ker f = \{0\}$; f is thus *injective* and, since it is assumed to be surjective, it is a bijection. Let $y \in F$. Put $x = f^{-1}(y)$; then $y = f(x)$ and $m \| x \| \leqslant \| f(x) \|$ may be written as $\| f^{-1}(y) \| \leqslant m^{-1} \| y \|$. Hence f^{-1} is continuous.

A.2.4 Equivalent norms

Two norms on a v.s. E are said to be equivalent if they define the same topology.

Theorem A.4

The following conditions are equivalent:
(a) The two norms $\| \ \|_1$ and $\| \ \|_2$ are equivalent.
(b) There are numbers $m > 0$ and $M > 0$ such that $m \| x \|_2 \leqslant \| x \|_1 \leqslant M \| x \|_2$ for all $x \in E$.
(c) The identity map of E (endowed with $\| \ \|_1$) to E (endowed with $\| \ \|_2$) is an isomorphism.

Proof. Plainly (a) $\Leftrightarrow$ (c), and (b) $\Leftrightarrow$ (c) follows from Theorem A.3. □

A.2.5 The case of finite-dimensional spaces

Theorem A.5

On a finite-dimensional space all norms are equivalent.

Proof. Every vector $x \in E$ may be written as $x = x_1 e_1 + \ldots + x_n e_n$ in a fixed basis (e_i) of E. It is clear that $\| x \| = | x_1 | + \ldots + | x_n |$ defines a norm. If $\| \ \|_1$ is another norm, $\| x \|_1 = \| \sum x_i e_i \|_1 \leqslant \sum | x_i | \ \| e_i \|_1 \leqslant M(| x_1 | + \ldots + | x_n |) = M \| x \|$, where $M = \sup\{\| e_1 \|_1, \ldots, \| e_n \|_1\}$. Hence $x \mapsto \| x \|_1$ is continuous for the topology of $\| \ \|$, for (see §A.1.1) we have $| \, \| x \|_1 - \| y \|_1 | \leqslant \| x - y \|_1 \leqslant M \| x - y \|$.

On the other hand, the unit sphere S of E, normed by $\| \ \|$, is closed and bounded, and hence compact, for the usual topology. As $x \mapsto \| x \|_1$ is continuous and positive on S, we must have $m = \inf_{x \in S} \| x \|_1 > 0$. Hence there are constants $m > 0$ and $M > 0$ such that $m \leqslant \| x \|_1 \leqslant M$ if $\| x \| = 1$, and the theorem follows from §A.2.4. □

Theorem A.6

Let E and F be finite-dimensional normed v.s. over the same field. Then every linear map $f: E \to F$ is continuous.

Proof

Let (e_i) (resp. (f_j)) be a basis for E (resp. F). Define on E the norm $\| x \| = \sum | x_i |$ as above, and on F the analogous norm $\| y \| = \sum | y_i |$. Then $\| f(x) \| = \| \sum x_i f(e_i) \| \leqslant \sum | x_i | . \| f(e_i) \| \leqslant A \sum | x_i | = A \| x \|$, where $A = \sup\{ \| f(e_1) \|, \ldots, \| f(e_n) \| \}$. Hence f is continuous with respect to the norms $\| \ \|$, and thus for all norms by Theorem A.5. □

A.3 MULTILINEAR CONTINUOUS MAPS

Definition A.1

Let $E_1, \ldots, E_n$ and F be v.s. over the same field. A map f: $E_1 \times \ldots \times E_n \to F$ is said to be multilinear (bilinear if $n = 2$) if the function $f(x_1, \ldots, x_n)$ is linear with respect to each of the variables x_i. (Note that this does not mean that f is a linear map of $E_1 \oplus \ldots \oplus E_n$ in F.)

Suppose further that $E_1, \ldots, E_n$ and f are normed. Then $E_1 \times \ldots \times E_n$ has the structure of a normed space (the direct sum of E_i) and it is permissible to speak of a continuous multilinear map f: $E_1 \times \ldots \times E_n \to f$. in view of this, there now follows a generalization of Theorem A.1.

Theorem A.7

The following conditions are equivalent:
(a) f is continuous at all points of $E_1 \times \ldots \times E_n$.
(b) f is continuous at the origin $(0, \ldots, 0)$ of $E_1 \times \ldots \times E_n$.
(c) $\|f(x_1, \ldots, x_n)\|$ is bounded on the product of the unit balls $\|x_i\| \leqslant 1$ of the E_i.
(d) $\|f(x_1, \ldots, x_n)\|$ is bounded on the product of the unit spheres $\|x_i\| = 1$ of the E_i.
(e) There is a constant M such that $\|f(x_1, \ldots, x_n)\| \leqslant M \|x_1\| \ldots \|x_n\|$ for all $x_1 \in E_1, \ldots, x_n \in E_n$.

Proof. Clearly (a) ⇒ (b), so we show that (b) ⇒ (c). There is a constant $\delta > 0$ such that $\|x_1\| + \ldots + \|x_n\| \leqslant \delta \Rightarrow \|f(x_1, \ldots, x_n)\| \leqslant 1$. Then $\|x_1\| \leqslant 1, \ldots, \|x_n\| \leqslant 1 \Rightarrow \|\delta x_1/n\| + \ldots + \|\delta x_n/n\| \leqslant \delta \Rightarrow \|f(\delta x_1/n, \ldots, \delta x_n/n)\| \leqslant 1 \Rightarrow \|f(x_1, \ldots, x_n)\| \leqslant (n/\delta)^n$.

Plainly (c) ⇒ (d), so we now show that (d) ⇒ (e). Let M be an upper bound for $\|f(x_1, \ldots, x_n)\|$ when $\|x_1\| = \ldots = \|x_n\| = 1$. If all the x_i are non-zero, the $x_i/\|x_i\|$ are of unit norm, and so $\|f(x_1/\|x_1\|, \ldots)\| \leqslant M$, whence $\|f(x_1, \ldots, x_n)\| \leqslant M\|x_1\| \ldots \|x_n\|$; this inequality plainly also holds if one of the x_i is zero.

Finally (e) ⇒ (a) follows from

$$\begin{aligned}\|f(x_1, \ldots, x_n) - f(x_1', \ldots, x_n')\| &= \|f(x_1 - x_1', x_2, \ldots, x_n) \\ &\quad + f(x_1', x_2 - x_2', x_3, \ldots, x_n) \\ &\quad + \ldots + f(x_1', \ldots, x_{n-1}', x_n - x_n')\| \\ &\leqslant M[\|x_1 - x_1'\| . \|x_2\| \ldots \|x_n\| \\ &\quad + \ldots + \|x_1'\| \ldots \|x_{n-1}'\| . \|x_n - x_n'\|].\end{aligned}$$ □

A.3.1 The space $(E_1, \ldots, E_n; F)$

The set of continuous multilinear maps f: $E_1 \times \ldots \times E_n \to F$, endowed in the natural way with operations of addition and scalar multiplication, is plainly a v.s. We shall denote it by $\mathscr{L}(E_1, \ldots, E_n; F)$. If $E_1 = \ldots = E_n = E$, we shall denote it by $\mathscr{L}^n(E; F)$.

Put $\|f\| = \sup\{\|f(x_1, \ldots, x_n)\| : \|x_1\| = \ldots = \|x_n\| = 1\}$. By the proof of (d) $\Rightarrow$ (e) of Theorem A.7 we have $\|f(x_1, \ldots, x_n)\| \leqslant \|f\| \, \|x_1\| \ldots \|x_n\|$ for all $x_1, \ldots, x_n$. We then verify, as in §A.2.1, that $f \rightsquigarrow \|f\|$ is indeed a norm on $\mathscr{L}(E_1, \ldots, E_n; F)$ and that, as in Theorem A.2, $\mathscr{L}(E_1, \ldots, E_n; F)$ with this norm is a Banach space if F is complete.

Example

Let E, F and G be normed v.s. over the same field. If $f \in \mathscr{L}(E; F)$ and $g \in \mathscr{L}(F; G)$, let $c(g, f)$ be the composite linear map $g \circ f$: $E \to G$. Clearly $c(g, f)$ is continuous. Thus c is a map of $\mathscr{L}(F; G) \times (E; F)$ to $\mathscr{L}(E; G)$. We know that it is bilinear and (see §A.2.2) that $\|g \circ f\| \leqslant \|g\| \, . \, \|f\|$; hence it is continuous and has norm $\|c\| \leqslant 1$.

A.4 CANONICAL ISOMORPHISMS

Theorem A.8

Let E be a normed v.s. over the field K. Then $\mathscr{L}(K; E)$ is canonically isomorphic to E.

Proof. Define a map $\hat{}\, : \mathscr{L}(K; E) \to E$ by $f \rightsquigarrow \hat{f} = f(1)$. Clearly $\hat{}$ is linear. It is injective, for $\hat{f} = 0 \Rightarrow f(1) = 0 \Rightarrow f(t) = tf(1) = 0$ for all $t \in K \Rightarrow f = 0$. It is surjective: if $x \in E$, set $f(t) = tx$, where $t \in K$; clearly $f \in \mathscr{L}(K; E)$ and $\hat{f} = x$. Lastly, it is continuous and $\|\hat{f}(t)\| = |t| \, . \, \|f\|$ shows that its norm $\|\hat{}\,\| = 1$. □

Theorem A.9

$\mathscr{L}(E; \mathscr{L}^k(E; F))$ is canonically isomorphic to $\mathscr{L}^{k+1}(E; F)$.

Proof. If $f \in \mathscr{L}(E; \mathscr{L}^k(E; F))$, define $\hat{f}$ by $\hat{f}(x_1, \ldots, x_{k+1}) = f(x_{k+1})(x_1, \ldots, x_k)$. Clearly $\hat{f} \in \mathscr{L}^{k+1}(E; F)$ and $\hat{}$ is a map of $\mathscr{L}(E; \mathscr{L}^k(E; F))$ to $\mathscr{L}^{k+1}(E; F)$. It is plain that $\hat{}$ is linear. It is injective, for $\hat{f} = 0 \Rightarrow f(x_{k+1})(x_1, \ldots, x_k) = 0$ for all $x_1, \ldots, x_{k+1} \Rightarrow f(x_{k+1}) = 0$ for all $x_{k+1} \Rightarrow f = 0$. It is surjective: if $g \in \mathscr{L}^{k+1}(E; F)$, define f: $E \to \mathscr{L}^k(E; F)$ by $x \rightsquigarrow g(x, \ldots)$; clearly $f \in \mathscr{L}(E; \mathscr{L}^k(E; F))$ and $\hat{f} = g$. Lastly, $\hat{}$ is continuous, for

$$\|\hat{f}(x_1, \ldots, x_{k+1})\| = \|f(x_{k+1})(x_1, \ldots, x_k)\| \leqslant \|f(x_{k+1})\| \, . \, \|x_1\| \ldots \|x_k\|$$
$$\leqslant \|f\| \, \|x_{k+1}\| \, \|x_1\| \ldots \|x_k\|$$

shows that $\|\hat{f}\| \leqslant \|f\|$, and so $\|\hat{}\,\| \leqslant 1$. (As an exercise, it should be proved that $\|\hat{}\,\| = 1$.) □

Appendix B

The Banach fixed-point theorem

B.1 FIXED POINTS

By a fixed point of a map f of a set E to itself is meant a point $a \in E$ such that $f(a) = a$.

Here is one of the reasons for the importance of this concept. If E is a v.s. and g a map of E to E, to solve the equation $g(x) = b$ for x, where $b \in E$ is given, amounts to finding the fixed point (or points) of $f(x) = g(x) + (x - b)$.

B.2 CONTRACTIONS

Let E be a metric space with metric d. A map f of a subset U of E in E is called a contraction if there is a real number k, $0 \leqslant k < 1$, such that $\mathrm{d}(f(x), f(y)) \leqslant k\,\mathrm{d}(x, y)$ for all $x, y \in U$. We call k the contraction ratio. Observe that a contraction is uniformly continuous.

B.3 BANACH'S THEOREM (THE CONTRACTION MAPPING PRINCIPLE)

Theorem B.1

Let E be a complete metric space, let U be an open subset of E, and let $f\colon U \to E$. Suppose that:

(a) f is a contraction with ratio k.

(b) There exists $u \in U$ whose distance $\mathrm{d}[u, E\backslash U]$ to the complement of U exceeds a fixed number $M > 0$.

(c) $\mathrm{d}(u, f(u)) < M(1 - k)$.

Then f has a unique fixed point $a \in U$ and $\mathrm{d}(u, a) < M$.

Proof. Plainly $f(u) \in M$, for $\mathrm{d}(u, f(u)) < M(1 - k) < M < \mathrm{d}(u, E/U)$. Define u_n recursively: $u_0 = u, \ldots, u_{n+1} = f(u_n)$. For this to make sense we must show that $u_n \in U$. We already know this for u_0 and u_1. Suppose it has been established for $u_0, u_1, \ldots, u_n$. We have $\mathrm{d}(u, f(u_n)) \leqslant \mathrm{d}(u, u_1) + \ldots + \mathrm{d}(u_n, f(u_n))$. But clearly:

$$\mathrm{d}(u_p, f(u_p)) = \mathrm{d}(f^p(u), f^p(u_1)) \leqslant k^p M(1 - k) \quad \text{for} \quad 0 \leqslant p \leqslant n$$

Thus:

$$d(u, f(u_n)) \leqslant M(1-k)(1+\ldots+k^n) = M(1-k^{n+1}) < M.$$

Hence $f(u_n) = u_{n+1} \in U$, and the property is established by induction. The same reasoning also shows that:

$$d(u_p, u_{p+q}) \leqslant M(1-k)(k^p+\ldots+k^{p+q-1}) < Mk^p.$$

The sequence u_n is thus a Cauchy sequence in U, which converges to a limit a, since E is complete. This limit lies in U, for the last inequality shows, on letting $q \to +\infty$, that $d(u_p, a) \leqslant Mk^p < M$. Also, letting $p \to +\infty$, we obtain $d(u, a) < M$.

We now show that a is a fixed point. Since f is continuous, $u_{n+1} = f(u_n)$ gives $a = f(a)$ on letting $n \to \infty$.

Let b be a second fixed point. Then $d(a, b) = d(f(a), f(b)) \leqslant d(a, b)$, and so $d(a, b) = 0$. Hence $b = a$.

Remark

If $U = E$, hypotheses (b) and (c) are redundant, and any point u will do for the construction of the Cauchy sequence in the proof. We give two useful corollaries.

B.4 DEPENDENCE ON A PARAMETER

Corollary B.1

With the same notation of Theorem B.1, let T be a topological space and let $f\colon U \times T \to E$ be a map which satisfies the following hypotheses:
(a) each partial map $f_t : x \mapsto f(x, t)$, where $t \in T$, of U to E satisfies the hypotheses of Banach's theorem.
(b) For each $x \in U$, the partial map $t \mapsto f(x, t)$ is continuous.

Then if a_t is the unique fixed point of f_t, the map $t \mapsto a_t$ is continuous.

Proof. Let s, $t \in T$. We have:

$$d(a_t, a_s) = d(f_t(a_t), f_s(a_s)) \leqslant d(f_s(a_s), f_s(a_t)) + d(f_s(a_t), f_t(a_t))$$

$$\leqslant k\, d(a_t, a_s) + d(f_s(a_t), f_t(a_t)).$$

Thus $d(a_t, a_s) \leqslant (1-k)^{-1}\, d(f_s(a_t), f_t(a_t))$, which tends to zero as $s \to t$. □

B.5 WHEN AN ITERATE IS CONTRACTION

Corollary B.2

If f is a continuous map of a complete metric space to itself such that some iterate f^r is a contraction with ratio k, then f has a unique fixed point.

Proof. By Banach's theorem, f^r has a fixed point $a = \lim_{n\to\infty} f^{rn}(u)$. Since f is continuous, $f(a) = f(\lim \ldots) = \lim_{n\to\infty} f^{rn+1}(u)$. But $d(f^{nr}(f(u)), f^{nr}(u)) \leqslant k^n d(f(u), u)$ and so $\lim_{n\to\infty} f^{rn+1}(u) = \lim_{n\to\infty} f^{rn}(u) = a$.

If $f(b) = b$, then $f^r(b) = b$; and $b = a$ since f^r has only one fixed point. □

B.6 CONSTRUCTIVE CHARACTER OF BANACH'S THEOREM

As we have seen in Chapter 3 connection with the inverse function theorem, Banach's theorem enables us to prove existence and uniqueness of solutions of equations. But it also provides an effective procedure for the construction of a solution by successive approximations. The proof of Banach's theorem shows that the sequence $u_0 = u$, $u_1 = f(u), \ldots, u_n = f^n(u)$ converges to the solution a of $f(x) = x$. It also shows that the *'rate of convergence' of the approximate solutions u_n to the solution is polynomial in n: $d(a, u_n) \leqslant Mk^n$.*

For example, let f: $[a, b] \to \mathbf{R}$ be differentiable, with derivative such that $|f'(x)| \leqslant k < 1$ for some fixed k and for all $x \in [a, b]$. Also assume that f maps $[a, b]$ into itself. The mean-value theorem (Theorem 2.3) shows that $|f(x) - f(y)| \leqslant k|x - y|$ for all $x, y \in [a, b]$. Thus the function f is a contraction of $[a, b]$ into $[a, b]$, and Figures B.1 and B.2 show the successive approximations x_n of the solution of $f(x) = x$ in the cases $0 < f'(x) < 1$ and $-1 < f'(x) < 0$.

The scope of the method is more general than might appear. Suppose we wish to solve an equation of the form $F(x) = 0$, where F: $[a, b] \to \mathbf{R}$ is a differentiable function with $F(a) < 0$, $F(b) > 0$, and such that there are two positive constants k_1, k_2 such that $k_1 \leqslant F'(x) \leqslant k_2$ for all $x \in [a, b]$.

Setting $f(x) = x - tF(x)$, $t \neq 0$, we are led to look for solutions of $f(x) = x$. Since $f'(x) = 1 - tF'(x)$, we have $1 - tk_2 \leqslant f'(x) \leqslant 1 - tk_1$. The previous procedure will now apply, adjusting t so that $|1 - tk_2| < 1$, $|1 - tk_1| < 1$, for then f will be a contraction.

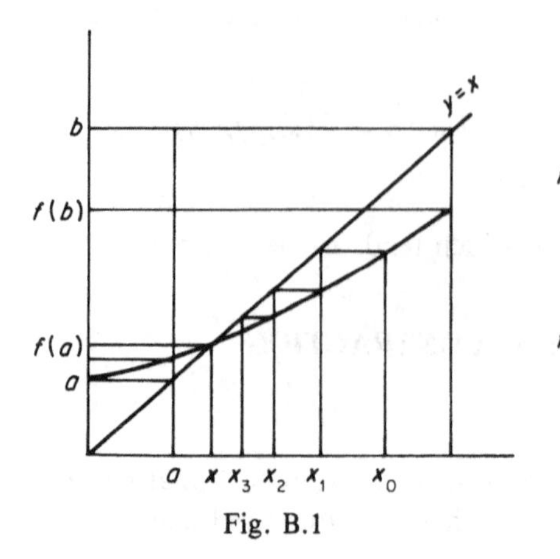

Fig. B.1

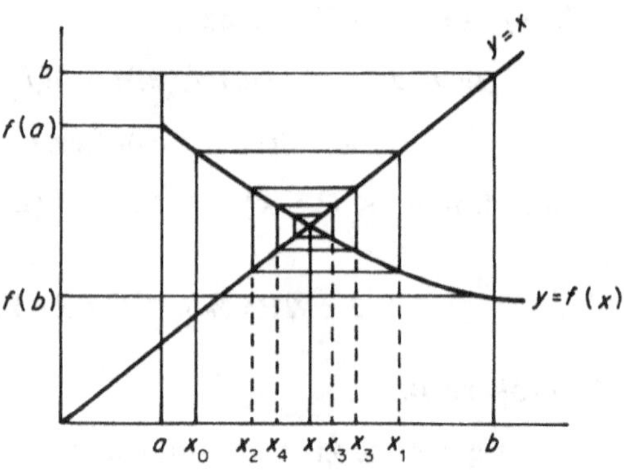

Fig. B.2

Remark

Suppose we are given a v.s. E and a map $f\colon E \to E$. We intend to solve the equation $f(x) = x$. If there is a norm on E with respect to which (a) E is a Banach space and (b) f is a contraction, then Banach's theorem provides a solution.

There may be several norms satisfying (a); in particular, this is so if E is finite-dimensional. In such a case, the conditions under which f is a contraction may depend on the norm chosen. This requires us to look for the 'most advantageous' norm in each particular case.

We can illustrate this by an example: Let $x = (x_1, \ldots, x_n) \in \mathbf{R}^n \rightsquigarrow f(x) = Lx + b \in \mathbf{R}^n$ be an affine map ($L \in \mathscr{L}(\mathbf{R}^n; \mathbf{R}^n)$ and $b \in \mathbf{R}^n$). Under what conditions is f a contraction?

First take the norm $\| x \|_1 = \sup_i | x_i |$. It is easy to see that:

$$\| f(x) - f(y) \|_1 \leqslant \sup_i \sum_{j=1}^{n} | L_{ij} | \, \| x - y \|_1$$

where (L_{ij}) is the matrix of L in the canonical basis. A sufficient condition for f to be a contraction is thus $\sum_{i=1}^{n} | L_{ij} | < 1$ for $i = 1, \ldots, n$.

Now take the norm $\| x \|_2 = \sum_{i=1}^{n} | x_i |$. Then:

$$\| f(x) - f(y) \|_2 \leqslant \sup_j \sum_{i=1}^{n} | L_{ij} | \, \| x - y \|_2$$

and a sufficient condition is:

$$\sum_{i=1}^{n} | L_{ij} | < 1 \quad \text{for} \quad j = 1, \ldots, n$$

Finally take the euclidean norm $\| x \| = \sqrt{\left(\sum_{i=1}^{n} x_i^2 \right)}$. Schwarz' inequality shows that $\| f(x) - f(y) \|^2 = \sum_i \left[\sum_j L_{ij}(x_j - y_j)^2 \right] \leqslant \sum_i \sum_j L_{ij}^2 \| x - y \|^2$. The condition that f should be a contraction is then $\sum_{ij} L_{ij}^2 < 1$.

Each of these conditions is *sufficient* for f to be a contraction; but none is necessary to ensure the convergence of the sequence $x, f(x), \ldots, f^n(x)$. The reader should construct examples in which one of these last three conditions is satisfied but the other two are not.

Appendix C

Newton's method

Banach's theorem in Appendix B gives an algorithm for finding the roots of an equation $f(x) = 0$. If we set $g(x) = f(x) + x$, we have seen that, under certain conditions, the sequence $u_0, u_1 = g(u_0), \ldots, u_{n+1} = g(u_n)$ converges to the unique root a and that the *rate of convergence is polynomial*: $\| a - u_n \| \leqslant Mk^n$, where M and k, $0 < k < 1$, are constants.

Newton's method for finding a root of the equation $f(x) = 0$ consists of replacing the curve $y = f(x)$ by its tangent at a point, the abscissa x_0 of which is an approximation to the desired root a. If $| a - x_0 | < \varepsilon$, the distance between the curve and its tangent is of order ε^2. The approximation x_1, defined by the linearized equation $f(x_0) + Df(x_0) . (x_1 - x_0) = 0$, is therefore a distance of order ε^2 from a (see Figure C.1).

Iteration of this procedure gives an exponentially convergent sequence of approximations: $| x_{n+1} - x_n | \leqslant C | x_n - x_{n-1} |^2$, $C =$ constant. The error $| a - x_n |$ of the $(n+1)$th approximation is thus of order $\varepsilon 2^n$.

We shall extend this method to Banach spaces.

Theorem C.1

Let E and F be Banach spaces, let B be a ball of radius $\delta > 0$ in E, and let f: $B \to F$ be a C^1 map. Assume that:

(a) There is a constant M such that $\| Df(x) - Df(y) \| \leqslant M \| x - y \|$ for all $x, y \in B$.

(b) There exists $x_0 \in B$ such that $Df(x_0)$ is an isomorphism of E on F and such that $\| f(x_0) \|$ is small enough.

Then $x_{n+1} = x_n - [Df(x_n)]^{-1} f(x_n)$ exists, and the sequence x_n converges to the unique solution a of $f(x) = 0$. Moreover, $\| a - x_n \|$ is of order $\varepsilon 2^n$, where ε is a constant, $(0 < \varepsilon < 1)$.

Proof. Let $x_0 \in B$ be an approximation to the desired root a. Set $y_0 = f(x_0)$; we have to solve for r the equation $f(x_0 + r) = 0$, which may also be written as $f(x_0 + r) - f(x_0) = -y_0$. Suppose we know that $x_0 + r \in B$; then $f(x_0) = f(x_0 + r) - Df(x_0 + r) . r + R(x_0, r)$, where $R = O(\| r \|)$. Then r would satisfy the equation $-y_0 = Df(x_0 + r) . r - R(x_0, r)$.

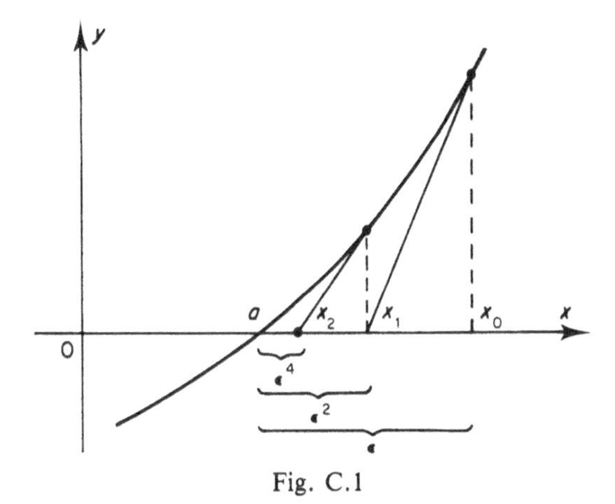

Fig. C.1

Since Df is continuous and $Df(x_0)$ is an isomorphism, $Df(x_0+r)$ is also an isomorphism (Lemma 3.2). Thus the map ϕ: $r \mapsto -[Df(x_0+r)]^{-1}$. $[y_0 - R(x_0, r)]$ is well defined if $\| r \|$ is small enough.

We shall prove that:

(1) ϕ is a contraction of the ball $B(x_0, \delta')$ into itself, if the radius δ' is small enough.

(2) The algorithm for the construction of the fixed point of this contraction given by Banach's theorem is the same as that given in the statement of the theorem.

We begin by estimating $g(r) = [Df(x_0+r)]^{-1}$. The identity $A^{-1} - B^{-1} = A^{-1}(B-A)B^{-1}$ between linear isomorphisms, and hypothesis (a) of the theorem, show that $\| g(r) - g(r') \| \leqslant M \| r - r' \| \| g(r) \| \| g(r') \|$. With $C = \| g(0) \|$, this gives $\| g(r) \| \leqslant \| g(r) - g(0) \| + \| g(0) \| \leqslant CM \| r \| \| g(r) \| + C$, whence $\| g(r) \| \leqslant C[1 - CM \| r \|]^{-1}$.

Thus if $\| r \| \leqslant \delta' < \delta$ and $\| r \| < (2MC)^{-1}$, then $g(r)$ exists and $\| g(r) \| \leqslant 2C$.

Moreover, if $r, r' \in B(x_0, \delta')$, we have $\| \phi(r) - \phi(r') \| \leqslant \| g(r) \| \| R(x_0, r) - R(x_0, r') \| + \| g(r) - g(r') \| \| y_0 - R(x_0, r') \|$. But by Corollary 2.3 and hypothesis (a) of the theorem, we have:

$$\| R(x_0, r) - R(x, r') \| = \| f(x_0 + r') - f(x_0 + r) + Df(x_0+r) - Df(x_0+r')r' \|$$
$$\leqslant \| f(x_0+r') - f(x_0+r) - Df(x_0+r).(r'-r) \| + \| [Df(x_0+r) - Df(x_0+r')]r' \|$$
$$\leqslant M[\| r' - r \|^2 + \| r' \| \| r' - r \|].$$

Using the above results and hypothesis (a) of the theorem, we obtain:

$$\| \phi(r) - \phi(r') \| \leqslant 2CM[\| r' - r \|^2 + \| r' \| \| r' - r \|] + 4C^2M(\| y_0 \| + \tfrac{1}{2}M \| r \|^2) \| r' - r \|,$$

that is, $\| \phi(r) - \phi(r') \| \leqslant K \| r - r' \|$, where K is a positive constant, independent of r and r', and less than 1 if δ' and $\| y_0 \|$ are small enough.

Moreover, $\| \phi(r) \| \leqslant \| \phi(r) - \phi(0) \| + \| \phi(0) \| \leqslant K \| r \| + C \| y_0 \|$, and so $\| \phi(r) \| \leqslant \delta'$ if $C \| y_0 \| \leqslant (1 - K)\delta'$.

To sum up, choosing $\| y_0 \| = \| f(x_0) \|$, and therefore δ', small enough, ϕ is a contraction of $B(x_0, \delta')$ into itself. By Banach's theorem, ϕ has a unique fixed point $a \in B(x_0, \delta')$, which is the limit of the sequence defined by $r_0 = 0$:

$$\begin{aligned} r_{n+1} = \phi(r_n) &= -\,[Df(x_0 + r_n)]^{-1}[y_0 - R(x_0, r_n)] \\ &= -\,[Df(x_0 + r_n)]^{-1}[f(x_0 + r_n) - Df(x_0 + r_n) \,.\, r_n] \\ &= r_n - [Df(x_0 + r_n)]^{-1} f(x_0 + r_n). \end{aligned}$$

Setting $x_n = x_0 + r_n$ we obtain the recurrence relation stated in the theorem.

Let us estimate the rate of convergence. By Corollary 2.3 and hypothesis (a) of the theorem, we have $\| f(x_n) - f(x_{n-1}) - Df(x_n)(x_n - x_{n-1}) \| \leqslant M \| x_n - x_{n-1} \|^2$. Thus $\| x_{n+1} - x_n \| \leqslant D \| x_n - x_{n-1} \|^2$, where D is an upper bound of the $M \| [Df(x_n)]^{-1} \|$. It follows that:

$$\begin{aligned} \| x_{n+1} - x_n \| \leqslant D \| x_n - x_{n-1} \|^2 \leqslant \ldots &\leqslant D^{1+2+\ldots+2^{n-1}} \| x_1 - x_0 \|^{2^n} \\ &= D^{-1}(D \| x_1 - x_0 \|)^{2^n}. \end{aligned}$$

Thus if we put $\varepsilon = D \| x_1 - x_0 \| = D \| [Df(x_0)]^{-1} \| \, \| f(x_0) \|$ and $a = \lim_{n \to \infty} x_n$, we obtain:

$$\| a - x_n \| \leqslant \sum_{j=0}^{\infty} \| x_{n+j+1} - x_{n+j} \| \leqslant D(\varepsilon^{2^n} + \varepsilon^{2^{n+1}} + \ldots) = O(\varepsilon^{2^n}). \quad \square$$

Appendix D

Global inverse function theorems

A C^1 map f: $\mathbf{R} \to \mathbf{R}$ whose derivative never vanishes is a diffeomorphism of a neighbourhood of each point on the image under f of this neighbourhood. But without supplementary conditions it is not necessarily a diffeomorphism of $\mathbf{R}$ on to $\mathbf{R}$ (think of $\tan^{-1}x$). Such conditions may be of a global character; for example: $\lim_{|t| \to \infty} |f(t)| = \infty$. They may also have a local nature; for example: there is a number k such that $f'(x) > k$ for all x. It is this latter condition that we are going to adapt to Banach spaces. Notice that it may be written in two ways, each of which will give rise to a global theorem:

(a) There is a number $k > 0$ such that $[f(y) - f(x)](y - x) \geqslant k(y - x)^2$.

(b) There is a number $k > 0$ such that $|[Df(x)]^{-1}| < 1/k$ for all x.

D.1 STRICTLY MONOTONE MAPS

In this section H will be a real Hilbert space, with scalar product $\langle , \rangle$. With minor changes, which are left to the reader, the following results may be extended to complex Hilbert spaces.

Definition D.1

Let us generalize condition (a) of the introduction. A map f: $H \to H$ is called strictly monotone if there is a number $k > 0$ such that $\langle f(y) - f(x), y - x\rangle \geqslant k \| x - y \|^2$ for all $x, y \in H$.

If H were a complex Hilbert space, the left-hand side could be replaced by its modulus or by its real part.

Notice that Schwarz' inequality gives:

$$\| f(y) - f(x) \| \geqslant k \| x - y \|. \tag{D.1}$$

Theorem D.1

A strictly monotone C^1 map is a C^1 diffeomorphism of H on to H. The proof follows from the next two lemmas.

Lemma D.1

The derivative of f satisifes:

$$\langle Df(x)y, y\rangle \geqslant k \| y \|^2 \text{ for all } x, y \in H. \tag{D.2}$$

Proof. For all real t and for all $x, y \in H$ we have $\langle f(x+ty) - f(x), ty\rangle \geqslant kt^2 \| y \|^2$. Dividing both sides by $t^2 \neq 0$, letting t tend to zero, and using the continuity of $u \rightsquigarrow \langle u, y\rangle$, we obtain equation (D.2). □

Note that, conversely, equation (D.2) implies strict monotonicity, for which we thus have a local criterion; for $\langle Df(x+ty)y, y\rangle \geqslant k \| y \|^2$ for all $t \in [0, 1]$ and all $x, y \in H$. Integrate both sides from 0 to 1 and use the fundamental theorem of integral calculus (Theorem 2.10) and property (d) of §2.6.4:

$$\langle f(x+y) - f(x), y\rangle = \langle \int_0^1 Df(x+ty)y \, dt, y\rangle$$

$$= \int_0^1 \langle Df(x+ty)y, y\rangle \, dt \geqslant k \| y \|^2.$$

D.1.1 The Lax-Milgram lemma

Lemma D.2

Let H be a real Hilbert space and let $L \in \mathscr{L}(H; H)$. Suppose there is a number $k > 0$ such that $\langle Lx, x\rangle \geqslant k \| x \|^2$ for all $x \in H$. Then L is an isomorphism and $\| L^{-1} \| \leqslant 1/k$.

Proof. Since $x = 0$ if $Lx = 0$, L is injective.

We shall prove that $L(H)$ is a closed subspace. Let $y_n = L(x_n)$ be a sequence which converges to y. Schwarz' inequality shows that $\| Lx \| \geqslant k \| x \|$ for all x, and so:

$$\| y_p - y_q \| = \| L(x_p - x_q) \| \geqslant k \| x_p - x_q \|$$

and x_n is a Cauchy sequence. Since H is complete and L is continuous, x_n converges to an element x and $L(x) = y$.

We now show that L is surjective. The projection theorem shows that it is enough to prove that the orthogonal complement A of the closed v.s. $L(H)$ is $\{0\}$. If $a \in A$, then $0 = \langle La, a\rangle \geqslant k \| a \|^2$, and so $a = 0$.

The linear map L thus has an inverse L^{-1}. Putting $Lx = y$, the inequality $\| Lx \| \geqslant k \| x \|$ may be written as $\| y \| \geqslant k \| L^{-1}y \|$. Hence L^{-1} is bounded and $\| L^{-1} \| \leqslant 1/k$.

D.1.2 Proof of Theorem D.1

Equation (D.1) shows that f is injective. Lemmas D.1 and D.2 show that, for all x, $Df(x)$ is an isomorphism. By the invariance of domain theorem (See Corollary 3.1), f is thus a C^1 diffeomorphism of E on the open set $F = f(E)$.

It remains to show that f is surjective. To do this it is enough to prove that $f(E)$ is closed, or that every boundary point b of $f(E)$ belongs to $f(E)$.

There is a path, that is, a continuous curve γ: $[0,1] \to H$, of class C^1 on $[0,1[$, with end $\gamma(1) = b$, such that $\gamma(t) \in f(H)$ for $0 \leqslant t < 1$. Since $f: E \to f(H)$ is a C^1 diffeomorphism, $\tilde{\gamma} = f^{-1} \circ \gamma|_{[0,1[} : [0,1[\to H$ is of class C^1 and $f \circ \tilde{\gamma}(t) = \gamma(t)$ for $0 \leqslant t < 1$. If t_n is an increasing sequence of numbers tending to 1, equation (D.1) shows that:

$$\| \gamma(t_p) - \gamma(t_q) \| = \| f[\tilde{\gamma}(t_p)] - f[\tilde{\gamma}(t_q)] \| \geqslant k \| \tilde{\gamma}(t_p) - \tilde{\gamma}(t_q) \|.$$

Thus $\tilde{\gamma}(t_n)$ is a Cauchy sequence, which converges to a point x. The continuity of f implies that $f(x) = \lim_{n \to \infty} f[\tilde{\gamma}(t_n)] = \gamma(1) = b$, and b therefore belongs to $f(H)$. □

D.2 THE HADAMARD–LÉVY THEOREM

In generalizing condition (b) of the introduction, we shall obtain a result due to Jacques Hadamard (1906) and Paul Lévy (1920).

Theorem D.2

Let E and F be Banach spaces, and let f: $E \to F$ be a C^1 map. Suppose that for all $x \in E$, $Df(x) \in \mathscr{L}(E; F)$ is an isomorphism and that there is a constant $A > 0$ such that $\| [Df(x)]^{-1} \| < A$. Then f is a diffeomorphism of E on F. There are several stages of the proof.

Lemma D.3

Let $x \in E$, $y = f(x)$, ϕ: $Z \to F$ a continuous map of a connected topological space Z to F. Suppose there is a point $z \in Z$ and a continuous map $\phi' : Z \to E$ such that $\phi(z) = y$, $\phi'(z) = x$, $f \circ \phi' = \phi$ (we say that ϕ' is a *lifting* of ϕ). Then ϕ' is unique.

Proof. Let ϕ'' be a map with the same properties as ϕ'. The space Z is the disjoint union of the set A of points at which ϕ' and ϕ'' have the same value, and of its complement B. We shall prove that A and B are open. Since $z \in A$ and Z is connected, it will follow that $Z = A$.

If $a \in A$, $\phi'(a) = \phi''(a)$ belongs to an open set U whose image under f is an open neighbourhood of $\phi(a) = f[\phi'(a)]$, for f is a local diffeomorphism. Thus $\phi'^{-1}(U) \cap \phi''^{-1}(U)$ is an open neighbourhood of a contained in A.

If $b \in B$, $\phi'(b)$ (resp. $\phi''(b)$) belongs to an open set U' (resp. U'') whose image under f is an open neighbourhood of $\phi(b) = f[\phi'(b)] = f[\phi''(b)]$. Thus $\phi'^{-1}(U') \cap \phi''^{-1}(U'')$ is an open neighbourhood of b contained in B. □

Lemma D.4

Let $x \in E$, $y = f(x)$, and z an arbitrary point of F. Then there is a C^1 path $\tilde{\gamma}$: $[0, 1] \to E$ with initial point $\tilde{\gamma}(0) = x$, which is a lifting of the line segment γ: $[0, 1] \to F$ defined by $\gamma(t) = (1 - t)y + tz$. In particular, f is surjective.

Proof. Let A be the set of those $a \in [0, 1]$ such that $\gamma|_{[0, a[}$ can be lifted up along a C^1 path $[0, a[\ \to E$ with initial point x.

We shall prove that A is non-empty. By the inverse function theorem, there is an open neighbourhood U of x whose image under f is an open neighbourhood V of y. Thus $\gamma^{-1}(V)$ contains an interval $[0, s[$ and $(f|_U)^{-1}$ is of class C^1, equals x when $t = 0$ and lifts $\gamma|_{[0, s[}$.

If $a, a' \in A$ and $a < a'$, then by Lemma D.3 the lifting of $\gamma|_{[0, a'[}$ coincides on $[0, a[$ with that of $\gamma|_{[0, a[}$.

Let α be the supremum of A. We prove that $\alpha \in A$. With obvious notation, $f \circ \tilde{\gamma}(t) = \gamma(t)$ for $0 \leqslant t < \alpha$, and so $Df[\tilde{\gamma}(t)] \circ \tilde{\gamma}'(t) = \gamma'(t)$. The hypothesis on Df thus implies that $\| \tilde{\gamma}(t) \| \leqslant AM$, where M is an upper bound of the $\|\gamma'(t)\|$, $0 \leqslant t \leqslant 1$. If t_n is an increasing sequence of numbers tending to α, the mean-value theorem shows that;

$$\| \tilde{\gamma}(t_p) - \tilde{\gamma}(t_q) \| \leqslant AM | t_p - t_q |$$

The sequence $\tilde{\gamma}(t_n)$ is thus a Cauchy sequence and so converges to a point b. The continuity of f means that $f(b) = \lim f \circ \tilde{\gamma}(t_n) = \lim \gamma(t_n) = b$, and $\tilde{\gamma}(\alpha) = b$ is well-defined.

Let us prove, by *reductio ad absurdum*, that $\alpha = 1$; this will establish the lemma. There is an open neighbourhood U' of $\tilde{\gamma}(\alpha)$ the image of which under f is an open neighbourhood V' of $\gamma(\alpha)$. Since $\alpha < 1$, $\gamma^{-1}(V')$ contains an open interval $I =]\alpha - \varepsilon, \alpha + \varepsilon[$. Define $\sigma: I \to E$ by $(f|_{U'})^{-1}\gamma$, then set $\tilde{\gamma}_1(t) = \tilde{\gamma}(t)$ if $0 \leqslant t \leqslant \alpha$, $\tilde{\gamma}_1(t) = \sigma(t)$ if $\alpha \leqslant t < \alpha + \varepsilon$. Evidently $\tilde{\gamma}_1$ is of class C^1, $\tilde{\gamma}_1(0) = x$ and $\tilde{\gamma}_1$ is a lifting of $\gamma|_{[0, \alpha+\epsilon[}$. This contradicts the definition of α.

Definition D.2

A path in F (resp. E)—that is, a continuous map γ: $[0, 1] \to F$ (resp. E)— is called a circuit if its initial point $\gamma(0)$ coincides with its terminal point $\gamma(1)$.

Lemma D.5

A circuit γ in F with initial point $y = f(x)$ may be lifted up to a circuit $\tilde{\gamma}$ in E with initial point x. In particular, f is injective.

Proof. Making a translation if necessary, we may assume that $y = 0$. Let $Z = \{0 \leqslant t \leqslant 1\} \times \{0 \leqslant s \leqslant 1\}$. We shall prove that ϕ: $Z \to F$, defined by $\phi(t, s) = s\gamma(t)$, has a lifting $\tilde{\phi}: Z \to E$ such that $\tilde{\phi}(t, 0) = x$ for $0 \leqslant t \leqslant 1$, $\tilde{\phi}(0, 1) = \tilde{\phi}(1, 1) = x$. The lemma will then follow on taking $\tilde{\gamma}(t) = \tilde{\phi}(t, 1)$.

By Lemmas D.3 and D.4, each segment $[0, 1] \ni s \mapsto s\gamma(t)$ has a unique lifting $\tilde{\phi}_t(s)$ such that $\tilde{\phi}_t(0) = x$. Set $\tilde{\phi}(t,s) = \tilde{\phi}_t(s)$. It is enough to show that for each $s \in [0,1]$, $t \rightsquigarrow \tilde{\phi}(t,s)$ is a circuit with initial point and terminal point x.

To do this, consider the set A of $a \in [0,1]$ such that, for $0 \leqslant s \leqslant a$, $t \rightsquigarrow \tilde{\phi}(t,s)$ has the desired properties. *Mutatis mutandis*, the argument of Lemma D.4 shows that $A = [0,1]$. We shall prove, by *reductio ad absurdum*, that $\alpha = \sup A = 1$.

Each point $\tilde{\phi}(t,\alpha)$ is contained in an open neighbourhood U_t, whose image under f is an open neighbourhood V_t of $f[\tilde{\phi}(t,\alpha)] = \phi(t,\alpha) = \alpha\gamma(t)$. Since $[0,1]$ is compact, a finite number of these V_t cover the image of $t \rightsquigarrow \alpha\gamma(t)$; let V' be their union and U' the union of the corresponding U_t. Since $\alpha < 1$, V' contains the image of a circuit $t \rightsquigarrow (\alpha+\varepsilon)\gamma(t)$ for some $\varepsilon > 0$. Then $t \rightsquigarrow (f|_{U'})^{-1} \circ \phi(t, \alpha+\varepsilon)$ is a lifting of this circuit, which is itself a circuit with initial point x, and the uniqueness of liftings shows that it is simply $\tilde{\phi}(t,\alpha+\varepsilon)$. This gives the desired contradiction.

If x and x' are two points of E with the same image $y = f(x) = f(x')$, the image under f of the line segment which joins them is a circuit $\tilde{\gamma}$ with initial point y. By what we have shown, this circuit may be lifted up to a unique circuit with initial point x. As this lifting must also be the segment xx', we must have $x = x'$, and so f is injective. □

D.2.1 Proof of Theorem D.2

We know that f is a C^1 bijection of E on F. Since $Df(x)$ is invertible at each point, $Df^{-1} = [Df \circ f^{-1}]^{-1}$ is continuous and f^{-1} is of class C^1.

Remark

A posteriori f is a covering, that is, a surjection such that each $y \in F$ has an open neighbourhood V whose inverse image is the disjoint union of open sets U_i (only one, in the present case), f being a homeomorphism of each U_i on V. *A priori* this is not clear: it is the hypothesis $\| [D\phi]^{-1} \| \leqslant A$ which gives rise to the surjectivity.

Appendix E

Reduction of linear endomorphisms

Let E be a complex n-dimensional vector space, and let $\mathrm{End}(E) = \mathscr{L}(E; E)$. If $A, B \in \mathrm{End}(E)$ we shall simply write AB for the composite map $A \circ B$, and shall denote the identity map id_E by 1.

E.1 THE CAYLEY–HAMILTON THEOREM

E.1.1 The spectrum and the characteristic polynomial

The set of scalars a such that $A - a.1$ is not invertible is called the spectrum of $A \in \mathrm{End}(E)$. The following properties are equivalent:

(a) a belongs to the spectrum of A.
(b) The kernel of $A - a.1$ is non-trivial.
(c) The determinant $\det(A - a.1)$ is zero.

An eigenvalue of A is an element of the spectrum of A; it is a root of the polynomial of degree n: $k \rightsquigarrow \det(A - k.1)$, called the characteristic polynomial of A.

Since E is finite-dimensional, to each eigenvalue a there corresponds an eigenvector $e \neq 0 : A(e) = ae$.

Lemma E.1

If $A \in \mathrm{End}(E)$ has n distinct eigenvalues $a_1, \ldots, a_n$, with corresponding eigenvectors $e_1, \ldots, e_n$, then these vectors form a basis of E. Moreover, if p is the characteristic polynomial of A, then $p(A) = 0$.

Proof. Suppose that there are scalars c_i such that $\sum c_i e_i = 0$. Apply A, $A^2, \ldots, A^{n-1}$ successively to both sides of this equality. We obtain n relations which may be written in the matrix form:

$$\begin{pmatrix} 1 & & 1 \\ a_1 & & a_n \\ \vdots & & \vdots \\ a_1^{n-1} & & a_n^{n-1} \end{pmatrix} \begin{pmatrix} c_1 e_1 \\ \\ \vdots \\ c_n e_n \end{pmatrix} = 0.$$

The $n \times n$ matrix is a Vandermonde matrix with determinant $\prod_{i<j} (a_i - a_j) \neq 0$. It follows that $c_1e_1 = \ldots = c_ne_n = 0$, so that the c_i are zero and the e_i form a basis of E.

Since $A(e_i) = a_ie_i$, the matrix of A in this basis is the diagonal matrix $\text{diag}(a_i)$. The matrix of $p(A)$ is thus the diagonal matrix $\text{diag}[p(a_i)]$, which is zero since the a_i are roots of p. Hence $p(A) = 0$.

Lemma E.2

The set of all $A \in \text{End}(E)$ with n distinct eigenvectors is dense open subset of $\text{End}(E)$.

Proof. Choose a basis of E. This enables us to identify A with its matrix, and thus $\text{End}(E)$ with C^{n^2}.

Since the space E is complex, the characteristic polynomial p of A has n roots $a_1, \ldots, a_n$, possibly coincident. A root is multiple if, and only if, it is also a root of the derivative polynomial p'. Let $b_1, \ldots, b_{n-1}$ be the roots of p'. p and p' have a common root if, and only if:

$$0 = \prod_{s=1}^{n-1} \prod_{r=1}^{n} (a_r - b_s) = p(b_1) \ldots p(b_{n-1}).$$

This last expression is a symmetric polynomial in $b_1, \ldots, b_{n-1}$. It is thus a polynomial in the coefficients of p and p', that is, a polynomial $P(a_{11}, \ldots, a_{nn})$ in the coefficients of the matrix of A. We have therefore efined a polynomial map $f\colon A \rightsquigarrow P(a_{11}, \ldots)$ of $\text{End}(E) = C^{n^2}$ to C. It cannot vanish on an open subset of $\text{End}(E)$ without being identically zero. This is not the case: if the eigenvalues of A are all distinct, $f(A) \neq 0$. Hence $f(A) \neq 0$ on the complement U of $f^{-1}(0)$, which is a dense open subset of $\text{End}(E)$.

Remark

Every isomorphism can thus be approximated arbitrarily closely by a diagonalizable endomorphism, in view of Lemma E.1.

E.1.2 The Cayley–Hamilton theorem

Theorem E.1

Let p_A be the characteristic polynomial of $A \in \text{End}(E)$. Then $p_A(A) = 0$.

Proof. If A belongs to the open set U above, all its eigenvalues are distinct. By Lemma E.1 we then have $p_A(A) = 0$. Since the map $A \in \text{End}(E) \rightsquigarrow p_A(A) \in \text{End}(E)$ is continuous and is zero on the dense open subset U, it must vanish everywhere. □

E.2 REDUCTION

Lemma E.3

Suppose that the characteristic polynomial p of $A \in \mathrm{End}(E)$ is the product of two coprime polynomials p_1 and p_2. Then:
(a) E is the direct sum of the kernels E_1 and E_2 of $p_1(A)$ and $p_2(A)$. Moreover, E_1 and E_2 are invariant under A.
(b) The restriction of A to E_i, $i = 1, 2$, has characteristic polynomial p_i, up to a multiplicative constant.

Proof.
(a) *E_i, $i = 1, 2$, is invariant under A.* If $x \in E_i$, then $p_i(A)x = 0$. But $p_i(A)$ and A commute, and so $p_i(A)Ax = Ap_i(A)x = 0$, whence $Ax \in E_i$.

(b) *We show that $p_1(A)E \subset E_2$.* Any vector in $p_1(A)E$ may be written as $p_1(A)x$. Now by the Cayley–Hamilton theorem, $p_2(A)p_1(A) = p(A) = 0$, and so $p_2(A)p_1(A)x = 0$, whence $p_1(A)x \in E_2$. Similarly $p_2(A)x \in E_1$.

(c) *We prove that $E = E_1 + E_2$.* Since p_1 and p_2 are coprime, by Bezout's theorem there are polynomials q_1 and q_2 such that $p_1q_1 + p_2q_2 = 1$. Then $p_1(A)q_1(A) + p_2(A)q_2(A) = 1$. It follows that, if $x \in E$, $p_1(A)q_1(A)x + p_2(A)q_2(A)x = x$. Set $x_2 = p_1(A)q_1(A)x$, $x_1 = p_2(A)q_2(A)x$; we have $x = x_1 + x_2$ and, by (b), $x_i \in E_i$.

(d) *Let us show that $E = E_1 \oplus E_2$.* All we have to prove is that $E_1 \cap E_2 = \{0\}$. If $x \in E_1 \cap E_2$, $p_1(A)x = p_2(A)x = 0$. Since we have seen in (c) that $x = p_1(A)q_1(A)x + p_2(A)q_2(A)x$, and since two polynomials in A commute, we have $x = q_1(A)[p_1(A)x] + q_2(A)[p_2(A)x] = q_1(A).0 + q_2(A).0 = 0$.

(e) Lastly, we show that the restrictions A_1 and A_2 of A to E_1 and E_2 have characteristic polynomials p_1 and p_2, up to multiplicative constants.

Choose a basis B of E which is the union of a basis B_1 of E_1 and a basis B_2 of E_2. If, by abuse of notation, we denote the matrices of A, A_1 and A_2 in the bases B, B_1 and B_2 again by A, A_1 and A_2 respectively, we have:

$$A = \begin{pmatrix} A_1 & 0 \\ 0 & A_2 \end{pmatrix}.$$

This shows that the characteristic polynomial of A is the product of the characteristic polynomials p_1' and p_2' of A_1 and A_2. Thus $p = p_1'\, p_2' = p_1p_2$.

Let k be a root of p_1'; we shall see that it is also a root of p_1. To k corresponds an eigenvector $x \neq 0$ of the restriction A_1 of A to E_1. Thus $x \neq 0$, $x \in E_1$ and $Ax = kx$.

Divide the polynomial $p_1(t)$ by $t - k$: $p_1(t) = (t-k)q(t) + p_1(k)$. Replace t by A: $p_1(A) = (A - k.1)q(A) + p_1(k)1$, and apply both sides to x. We have $p_1(A)x = 0$, for $x \in E_1$; $(A - k.1)x = 0$ for $Ax = kx$; thus $p_1(k)x = 0$. As $x \neq 0$, this shows that k is a root of p_1.

Thus every root of p_1' (resp. p_2') is a root of p_1 (resp. p_2). Since p_1 and p_2 are coprime, they have no common roots, and as $p_1'p_2' = p_1p_2$, we see that the roots of p_i', $i = 1, 2$, are all the roots of p_i, of the same order. Hence $p_1' = cp_1$, $p_2' = c^{-1}p_2$, where c is a constant. □

The lemma may be extended by induction to the case where p is the product of m coprime polynomials. We should then have proved the following result invoked in §5.5.3:

Theorem E.2

Let $k_1, \ldots, k_m$ be the distinct roots of the characteristic polynominal p of $A \in \mathrm{End}(E)$: $p(k) = (k_1 - k)^{r_1} \ldots (k_m - k)^{r_m}$.

Then E is the direct sum of the kernels $E_1, \ldots, E_m$ of $(A - k_1 . 1)^{r_1}, \ldots, (A - k_m . 1)^{r_m}$. The E_i are invariant under A and the restriction of A to E_i has characteristic polynomial $(k_i - k)^{r_i}$. In particular, $\dim E_i = r_i$.

E.3 SURJECTIVITY OF THE EXPONENTIAL

We have seen in §5.2.3 that for all $X \in \mathrm{End}(E)$ the exponential $\exp X$ is invertible. Conversely, we are going to see that given any invertible map a of a *complex* finite-dimensional vector space E, there is at least one $X \in \mathrm{End}(E)$ such that $A = \exp X$ (and hence infinitely many since, for every integer n, $\exp(X + 2\pi i n . 1) = \exp(X)$).

First case. Suppose that A has characteristic polynomial $P(k) = (a - k)^r$. Since A is invertible, $a \neq 0$ and there exists $x \in C$ such that $a = e^x$. Put:

$$N = a^{-1}A - 1 \text{ and } B = \sum_{i=1}^{r=1} (-1)^{i-1}N^i/i$$

By the Cayley–Hamilton theorem, $N^r = 0$; hence $\exp B = 1 + N$ by the classical determination of $\exp[\log(1 + u)]$. Thus $\exp(x . 1 + B) = e^x \exp B = A$.

General case. We use the notation of Theorem E.2. The restriction A_i of A to E_i has characteristic polynomial $(k_i - k)^{r_i}$, and by the first case there exists $X_i \in \mathrm{End}(E_i)$ such that $\exp(X_i) = A_i$. Take a basis of E formed by the union of bases of $E_1, \ldots, E_n$. If

$$X = \begin{pmatrix} X_1 & & 0 \\ & \ddots & \\ 0 & & X_m \end{pmatrix}$$

then:

$$e^X = \begin{pmatrix} e^{X_1} & & \\ & \ddots & \\ 0 & & e^{X_m} \end{pmatrix} = \begin{pmatrix} A_1 & & \\ & \ddots & \\ 0 & & A_m \end{pmatrix} = A$$

and X has the required property.

Appendix F

Linear differential equations with periodic coefficients

Let E be a Banach space and $A: \mathbf{R} \to \mathrm{End}(E)$ a continuous function. We shall assume in all that follows that A has period $T > 0$: $A(t+T) = A(t)$ for all $t \in \mathbf{R}$.

We shall see the consequences of this hypothesis for the solutions of the linear differential equation $x'(t) = A(t)x(t)$, $x(t) \in E$. The question is not trivial, for although A has period T this is not necessarily so for the solutions. It is known to be false if A is constant (see §5.7).

F.1 THE MONODROMY OPERATORS

Theorem F.1

The resolvent $R(a, t)$ of the differential equation $x' = Ax$ satisfies $R(a+T, t+T) = R(a, t)$ for all t.

Proof. By Theorem 6.1 we know that $(\mathrm{d}/\mathrm{d}t)\ R(a, t) = A(t)R(a, t)$ and $R(a, a) = 1$ (the identity map of E). Since $A(t) = A(t+T)$, $t \rightsquigarrow R(a+T, t+T)$ satisfies the same differential equation and the same initial condition $R(a+T, t+T) = 1$. By uniqueness, $R(a+T, t+T) = R(a, t)$. □

As a consequence, the general solution of $x' = Ax$ may be written as $f(t) = R(a, t)f(a)$. In particular, $f(a+T) = R(a, a+T)f(a)$. For all the solutions of $x' = Ax$ to have period T, it is necessary that $C(a) = R(a, a+T)$ should reduce to 1. We call $C(a)$ a *monodromy operator* of $x' = Ax$. Note that if A is constant, $C(a) = \mathrm{e}^{TA}$ does not depend on a. This is not the general case.

Theorem F.2

All the monodromy operators are conjugate and thus they have the same eignevalues.

Proof. Theorem F.1 and Chasles' relation for the integral product show that $C(b) = R(b, b+T) = R(a+T,\ b+T)\ R(a, a+T)\ R(b, a) = R(a, b)\ C(a)\ R(a, b)^{-1}$.

The eigenvalues of $C(.)$ are called the *characteristic numbers* of $x' = Ax$. If E is finite-dimensional and if A is constant, the characteristic numbers are the e^{rT}, where the r are the eigenvalues of A. □

Theorem F.3

For all solutions of $x' = Ax$ to be of period T, it is necessary and sufficient that $a \mapsto C(a)$ should take the value 1.

Proof. We have seen that the condition is necessary. To show that it is sufficient, note that if $C(a) = 1$, the last proposition shows that $C(t) = 1$ for all t. As $f(t+T) = C(t)f(t)$, we certainly have $f(t+T) = f(t)$ for all t. □

We now examine whether there is at least one periodic solution.

Theorem F.4

To each eigenvalue k of $C(a)$ there corresponds a solution f of $x' = Ax$ such that $f(t+T) = kf(t)$ for all t. In particular, there is a non-zero solution of period T if, and only if, 1 is an eigenvalue of C. This theorem is due to G. Floquet (1883).

Proof. Let $u \neq 0$ be an eigenvector of $C(a)$ with eigenvalue k. Let f be the solution of $x' = Ax$ such that $f(a) = u$. Then $f(a+T) = C(a)f(a) = kf(a)$. It follows that, if we put $g(t) = f(t+T) - kf(t)$, g is the solution of $x' = Ax$ satisfying $g(a) = 0$. By uniqueness, $g(t) = 0$ for all t.

If $k = 1$, the above solution is thus of period T. Conversely, if f is a non-zero solution of period T, $f(a) = f(a+T) = C(a)f(a)$, and so $f(a)$ is an eigenvector of $C(a)$ with eigenvalue 1. □

We shall see that the characteristic numbers of $x' = Ax$ are invariant under linear changes of variable.

Theorem F.5

Let S: $\mathbf{R} \to \mathrm{GL}(E)$ be a C^1 function with the same period T as A. If we make the change of variable $y(t) = S(t)x(t)$ in the equation $x' = Ax$, we obtain a differential equation $y' = A_1 y$, whose monodromy operator $C_1(a)$ is similar to that, $C(a)$, of the original equation. This theorem is due to A. M. Liapounov (1892).

Proof. A simple calculation shows that $y' = A_1 y$, where:

$$A_1 = (S' + SA)S^{-1} \tag{F.1}$$

Thus as S', S and A are of period T, so is A_1 and we may speak of the monodromy operator $C_1(a) = R_1(a, a+T)$, where R_1 is the resolvent of $y' = A_1 y$.

The general solution of $y' = A_1 y$ is $y(t) = R_1(a,t)y(a) = R_1(a,t)S(a)x(a)$. As $x(t) = R(a,t)x(a)$, we also have $y(t) = S(t)x(t) = S(t)R(a,t)x(a)$. Combining these results, and since $x(a)$ is arbitrary, we find $R_1(a, t) = S(t)R(a, t)S(a)^{-1}$. Take $t = a+T$; since $S(a+T) = S(a)$ it follows that $C_1(a) = S(a)C(a)S(a)^{-1}$.

This shows that given two periodic linear differential equations with the same period T, it is not always possible to transform one to the other by a linear change of variable of period T.

We now examine whether it is possible to reduce our equation to an equation $y' = A_1 y$, where A_1 is constant.

F.2 LIAPOUNOV'S THEOREM

Theorem F.6

Let E be a complex finite-dimensional vector space, and let $A\colon \mathbf{R} \to \mathrm{End}(E)$ be continuous and of period $T > 0$. Then there exists at least one C^1 function S: $\mathbf{R} \to \mathrm{GL}(E)$ of period T such that the change of variable $y(t) = S(t)x(t)$ reduces the equation $x' = Ax$ to an equation $y' = A_1 y$, where A_1 is constant.

Proof. Let $C(a) = R(a, a+T)$ be a monodromy operator of $x' = Ax$. There exists at least one $B \in \mathrm{End}(E)$ such that $C(a) = \mathrm{e}^{TB}$ (Appendix E). Take $S(t) = \mathrm{e}^{tB}R(a,t)^{-1}$; in other words, set $x(t) = R(a,t)\mathrm{e}^{-tB}y(t)$. Equation (F.1) shows that y satisfies the equation $y' = By$. Evidently S is of class C^1. It remains to show that it is of period T. Since $\mathrm{e}^{TB} = R(a, a+T)$, Theorem F.1 and Chasles' relation for the integral product show that:

$$\begin{aligned} S(t+T) &= \mathrm{e}^{(t+T)B}R(t+T, a) = \mathrm{e}^{tB}R(a, a+T)R(t+T, a) \\ &= \mathrm{e}^{tB}R(t+T, a+T) = \mathrm{e}^{tB}R(t, a) = S(t). \end{aligned}$$ □

Although the explicit determination of S necessitates that of the resolvent and may involve the solution of the equation $x' = Ax$, important consequences may be drawn from this result.

F.2.1 Form of solutions

We known from §5.5.3 the form of the general solution of the constant-coefficient equation $y' = By$. If $\lambda_1, \ldots, \lambda_m$ are the distinct eigenvalues of B, this solution is a linear combination of vectors $\mathrm{e}^{t\lambda_i}t^k a$, where $a \in E$ and $k <$ multiplicity of λ_i. It follows that the general solution of $x' = Ax$ is a linear combination of vectors $\mathrm{e}^{t\lambda_i}t^k S(t)^{-1}a$, where $S(t)$ has period $T > 0$ (and is therefore bounded.).

Since $C(a) = e^{TB}$, the eigenvalues of $C(a)$—that is, the characteristic numbers—are the $e^{T\lambda_i}$. We call the λ_i the *characteristic exponents* of $x' = Ax$.

One of the numerous consequences of this result, due to Liapounov, is that for all the solutions of $x' = Ax$ to tend to zero as $t \to +\infty$ (resp. $-\infty$) it is necessary and sufficient that the real parts of the characteristic exponents be negative (resp. positive).

Bibliography

The reader will find more detailed accounts in the books by F. R. Gantmacher, and by N. Rouche and J. Mawhin.

Appendix G

Existence of solutions of differential equations and their dependence on initial data

We shall deduce Theorems 7.1 and 7.5 of Chapter 7 from the implicit function theorem by a method due to J. Robbin (On the existence theorem for differential equations, *Proc. Amer. Math. Soc.*, 1968, **19**, 1005–1006).

Theorem G.1

Let U be an open subset of a Banach space E, and let X be a C^1 vector field on U. Then given any $t_0 \in \mathbf{R}$ and any $x_0 \in U$, there are an open neighbourhood V of x_0 in U, an interval $]t_0 - a, t_0 + a[$ and a map ϕ: $]t_0 - a, t_0 + a[\times V \to U$ such that:
(a) ϕ is of class C^1.
(b) $\phi(t_0, x) = x$ for $x \in V$.
(c) $\mathrm{d}\phi/\mathrm{d}t(t, x) = X[\phi(t, x)]$ for $(t, x) \in]t_0 - a, t_0 + a[\times V$.

Proof. Without loss of generality and to simplify the notation, we may assume that $t_0 = 0$, that x_0 is the origin 0 in E and that U is the ball $B_{2r}(0)$ with centre 0 and radius $2r$. Set $U_0 = B_r(0)$ and $I = [-1, 1]$.

Recall that $C^0(I; E)$ (resp. $C^1(I; E)$) denotes the Banach space of C^0 (resp. C^1) maps of I to E supplied with the norm:

$$\|f\|_{C^0} = \sup_{t \in I} \|f(t)\| \text{ (resp. } \|f\|_{C^1} = \|f\|_{C^0} + \|f'\|_{C^0}).$$

The set $C_0^p(I; E)$ of those $f \in C^p(I; E)$, $p = 0$ or 1, such that $f(0) = 0$ is a closed subspace of $C^p(I; E)$; it is thus a Banach space. Let $C_0^p(I; U_0)$ be the subset of those $f \in C_0^p(I; E)$ with values in U_0. This is an open subset of $C_0^p(I; E)$ as the image $f(I)$ of each curve $f \in C_0^p(I; U)$ is compact and is thus a positive distance from the complement of U in E; this is therefore also so for each curve close enough to f. □

G.1 THE MAPPING F

If $x \in U_0$ and $f \in C_0^1(I; U_0)$ we have $\| x + f(t) \| < 2r$; thus $x + f(t) \in U$ for $t \in I$ and $X[s + f(t)]$ is defined. We may thus define a map F of the open subset $R \times U_0 \times C_0^1(I; U_0)$ of the Banach space $\mathbf{R} \oplus E \oplus C_0^1(I; F)$ to the Banach space $C^0(I; E)$ by $F(k, x, f)$: $t \rightsquigarrow \mathrm{d}f(t)/\mathrm{d}t + kX[x + f(t)]$.

G.2 THE MAP f IS OF CLASS C^1

By the results of §2.3 it is enough to show that the partial derivatives exist and are continuous.

The map F is differentiable with respect to k, and $D_1F(k, x, f)$ is the element of $\mathscr{L}(\mathbf{R}; C^0(I; E))$ which is canonically identified (by Theorem A.8) with the element $t \in I \rightsquigarrow -X[x + f(t)]$ of $C^0(I; E)$. Since f is continuous, $\{x + f(t): t \in I\}$ is a compact set on which the continuous function X is uniformly continuous. Given $\varepsilon > 0$, there is a $\delta > 0$ such that $\| X[x + f(t)] - X(y) \| < \varepsilon$ if $t \in I$ and $\| x + f(t) - y \| < \delta$. Take $x_1, x_2 \in U_0$ such that $\| x_1 - x_2 \| < \delta/2$ and $f_1, f_2 \in C_0^1(I; U_0)$ such that $\| f_1 - f_2 \|_{C^1} < \delta/2$. Hence $\| x_1 + f_1(t) - x_2 - f_2(t) \| < \delta$ for $t \in I$, and thus:

$$\| D_1F(k_1, x_1, f_1) - D_1F(k_2, x_2, f_2) \| = \sup_{t \in I} \| X[x_1 + f_1(t)] - X[x_2 + f_2(t)] \| < \varepsilon$$

This shows that D_1F is continuous.

Since X is differentiable, F is differentiable with respect to x and $D_2F(k, x, f) = -kDX[x + f]$. This is the element of $\mathscr{L}(E; C^0(I; E))$ which maps $y \in E$ to the element $t \in I \rightsquigarrow -kDX[x + f(t)]\,y$ of $C^0(I; E)$. Since DX is continuous, we may show, just as for D_1F, that D_2F is continuous.

Also, since X is differentiable, F is differentiable with respect to f: if $h \in C_0^1(I; E)$, $D_3F(k, x, F) \,.\, h$ is the element of $\mathscr{L}[C_0^1(I; E); C^0(I; E)]$ which maps h to the curve:

$$t \in I \rightsquigarrow \frac{\mathrm{d}h(t)}{\mathrm{d}t} - kDX[x + f(t)]\,h(t) \text{ in } C^0(I; E)$$

As for D_2F we see that D_3F is continuous.

G.3 THE MAP H

The last calculation shows that $D_3F(0,0,0) = \mathrm{d}/\mathrm{d}t$; in other words, $D_3F(0,0,0)$ maps $h \in C_0^1(I; E)$ to the curve $t \in I \rightsquigarrow \mathrm{d}h(t)/\mathrm{d}t \in E$ in $C^0(I; E)$. The continuous linear map $D_3F(0,0,0)$ is thus surjective: $g \in C^0(I; E)$ is the image of $t \in I \rightsquigarrow \int_0^t g(s)\,\mathrm{d}s$. It is also injective: if $h'(t) = 0$ for $t \in I$, then $h(t) = \text{constant} = h(0) = 0$. Hence $D_3F(0,0,0)$ is an isomorphism of $C_0^1(I; E)$ on $C^0(I; E)$.

As, on the other hand, $F(0,0,0) = 0$, Corollary 3.2 of the implicit function theorem shows that we may find a neighbourhood $]-a, a[\times V$ of $(0,0)$ in

$\mathbf{R} \times U_0$ and a C^1 map H of this neighbourhood in $C_0^1(I; U_0)$ such that $F[k, x, H(k, x)] = 0$ for $(k, x) \in]-a, a[\times V$; that is:

$$\frac{\mathrm{d}}{\mathrm{d}t}[H(k, (x)(t)] = kX[x + H(k, x)(t)] \quad \text{if} \quad t \in I \tag{G.1}$$

G.4 THE MAP ϕ

Since $x \in V \subset U_0$ and $H(k, x)(t) \in U_0$, then $\phi(t, x) = x + H(k, x)(t/k)$, where $-a \leqslant k \leqslant a$, satisfies $\| \phi(t, x) \| < 2r$ and thus defines a map of $]-k, k[\times V$ to U.

By the definition of H, ϕ is of class C^1 in t and:

$$\frac{\mathrm{d}\phi}{\mathrm{d}t}(t, x) = \frac{\mathrm{d}}{\mathrm{d}t}[H(k, x)(t/k)] = k^{-1}.kX[x + H(k, x)(t/k)] = X[\phi(t, x)] .$$

Moreover, since $H(k, x) \in C_0^1(I; U_0)$, we have $H(k, x)(0) = 0$ and hence $\phi(0, x) = x$.

We have thus exhibited, for each $k \in [-a, a]$, an integral curve of X with initial point x and defined on $[-k, k]$. Any two of these curves then coincide on the intersection of their domains of definition, by uniqueness. Hence $H(k, x)(t/k)$ does not depend on k. If $t \neq 0$ it follows that:

$$\phi(t, x) = x + H(t, x)(1) \tag{G.2}$$

If $t = 0$, equation (G.1) shows that $\mathrm{d}[H(0, x)(t)]/\mathrm{d}t = 0$ for $t \in I$; thus $H(0, x)(t) = \text{constant} = H(0, x)(0) = 0$, and so $H(0, x)(1) = 0$. Equation (G.2) then holds also for $t = 0$ since $\phi(0, x) = x$. Since H is of class C^1 in (t, x), equation (G.2) shows that the same is true of ϕ.

Appendix H

Simplicity of SO(3)

We recall that SO(n) is the subgroup of O(n) formed by those elements a with det $(a) = +1$.

We have seen in §9.1.6 that O(3) is a C^∞ submanifold of End($\mathbf{R}^3$) = $\mathbf{R}^9$. We shall use the inverse function theorem (Theorem 9.12) to show that $G = \mathrm{SO}(3)$ is a simple group.

H.1 PRELIMINARIES

Proposition H.1

$G = \mathrm{SO}(3)$ is the connected component of $I = \mathrm{id}_{\mathbf{R}^3}$ in the submanifold O(3) endowed with the topology induced by $\mathbf{R}^9$.

Proof. The continuous map $\det : \mathrm{O}(3) \to \mathbf{R}$ has only the two values, $+1$ and -1. Thus O(3) has at least two connected components, contained in $\det^{-1}(1) = G$ and $\det^{-1}(-1)$.

It is accordingly enough to show that G is connected. An element $a \in G$ is a rotation of angle $\hat{a}$ about an axis. If $r(t)$ denotes the rotation of angle $t\hat{a}$ about the same axis, then $t \in [0, 1] \rightsquigarrow r(t) \in G$ is an arc with origin I and end-point a.

Proposition H.2

We may define a scalar product on End($\mathbf{R}^n$) by writing $\langle c, d \rangle = \mathrm{trace}(c \circ {}^t d)$, where ${}^t d$ is the transpose of d.

Moreover, if $s \in \mathrm{O}(\mathbf{R}^n)$, then $\langle s \circ c, s \circ d \rangle = \langle c, d \rangle$ and $\langle s \circ c \circ s^{-1}, s \circ d \circ s^{-1} \rangle = [c, d]$.

Proof. The first part is immediate. The second follows from ${}^t s = s^{-1}$ and from the fact that two similar endomorphisms have the same trace. □

Corollary H.1

Let $v \in \mathrm{End}(\mathbf{R}^n)$ and $a, b \in \mathrm{SO}(n)$. Then $2v = a \circ v \circ a^{-1} + b \circ v \circ b^{-1}$ implies that $v \circ a = a \circ v$, $v \circ b = b \circ v$.

Proof. We have $2\|v\|^2 = 2\langle v, v\rangle = \langle a\circ v\circ a^{-1}, v\rangle + \langle b\circ v\circ b^{-1}, v\rangle$. Use Proposition H.2 and Schwarz' inequality: $\langle a\circ v\circ a^{-1}, v\rangle \leqslant \|v\|^2$, with equality only if $a\circ v\circ a^{-1} = v$; and similarly with a replaced by b. The first inequality thus gives $a\circ v\circ a^{-1} = b\circ v\circ b^{-1} = v$.

H.2 SIMPLICITY OF SO(3)

Theorem H.1

Any invariant subgroup $H \neq I$ of $G = \mathrm{SO}(3)$ coincides with G.

Proof. Take $a \in H\backslash\{I\}$; this is a rotation around an axis D. If X is a non-zero vector along D, then $a(X) = X$.

Let r be a rotation which sends X to an orthogonal vector $Y = r(X)$. Then:

$$b = r\circ a\circ r^{-1} \in H \quad \text{and} \quad b(Y) = Y$$

Consider the map ϕ: $G \to G$ defined by $\phi(g) = g\circ a\circ g^{-1}\circ a^{-1}\circ g\circ b\circ g^{-1}\circ b^{-1}$. Plainly ϕ is C^∞, $\phi(I) = I$, and $\phi(G) \subset H$. We shall show that ϕ is a local diffeomorphism of G in the neighbourhood of I. By the inverse function theorem (Theorem 9.12), it is enough to prove that T_I is injective.

Let $v \in T_I G$. Take a curve γ traced on G, with origin I and satisfying $\gamma'(0) = v$. Its image under ϕ is $\phi\circ\gamma(t) = \gamma(t)\circ a\circ\gamma(t)^{-1}\circ a^{-1}\circ\gamma(t)\circ b\circ\gamma(t)^{-1}\circ b^{-1}$. By Theorem 4.5 and Leibniz' rule, $(\phi\circ\gamma)'(0) = 2v - a\circ v\circ a^{-1} - b\circ v b^{-1}$. Suppose that $v \in \operatorname{Ker} T_I\phi$. Then $2v = a\circ v\circ a^{-1} + b\circ v\circ b^{-1}$ and the corollary above shows that $v\circ a = a\circ v$, $v\circ b = b\circ v$. Now $a(X) = X$, $b(Y) = Y$; hence $a[v(X)] = v(X)$ and $b[v(Y)] = v(Y)$. But as a and b do not reduce to I, their only invariant vectors are those along their axis of rotation. Thus $v(X)$ is proportional to X and $v(Y)$ to Y. Since v is antisymmetric (see Example (c) in §9.2.2), $v(X) = v(Y) = 0$. As X and Y are orthogonal, this implies that $v = 0$. Thus $T_I\phi$ is injective and so bijective.

By the inverse function theorem, it now follows that there exists an open neighbourhood V of I such that $\phi(V)$ is an open neighbourhood of I. We have also seen that $\phi(V) \subset \phi(G) \subset H$.

We next prove that H is open and closed in G. Since G is connected (Proposition H.1) this will show that $G = H$. As H is a group, and left-translation L_h by $h \in H$ is a diffeomorphism of G (see §9.3.2), $L_h\phi(V)$ is an open neighbourhood of h contained in H. This left-translation by an element $h \in \mathrm{SO}(3)$ also preserves the euclidean distance defined on $\mathrm{End}(\mathbf{R}^3)$ in Proposition H.2, and so we see that H is closed. □

Note

A similar method enables us to establish the simplicity of the classical groups, such as $\mathrm{SO}(2k+1)$, $k \neq 0$.

Bibliography

Abraham, R., and Marsden, J. *Foundations of Mechanics*, 2nd ed., Benjamin, New York, 1978.

Arnold, V., *Equations Différentielles Ordinaires*, Mir, Moscow, 1974.

Cartan, H. *Cours de Calcul Différentiel*, 2nd ed., Hermann, Paris, 1977.

Coddington, E. A., and Levinson, N. *Theory of Ordinary Differential Equations*, McGraw Hill, New York, 1955.

Dieudonné, J. *Éléments d'Analyse*, Vol. 1: *Fondements de l'Analyse Moderne*, Gauthier-Villars, Paris, 1968.

Gantmacher, F. R. *The Theory of Matrices*, Chelsea, New York, 1959.

Hirsch, M. W., and Smale, S. *Differential Equations, Dynamical Systems, and Linear Algebra*, Academic Press, New York, 1974.

Malliavin, P. *Géométrie Différentielle Intrinsèque*, Hermann, Paris, 1972.

Milnor, J. *Topology from the Differentiable Point of View*, The University Press of Virginia, Charlottesville, 1965.

Nelson, E. *Topics in Dynamics*, Vol. 1: *Flows*, Princeton University Press, Princeton, 1969.

Rouche, N., and Mawhin, J. *Équations Différentielles Ordinaires*, Masson, Paris, 1973.

Young, L. C. *Lectures on the Calculus of Variations and Optimal Control Theory*, 2nd ed., Chelsea, New York, 1980.

Index